SIX ROADS FROM NEWTON

SIX ROADS FROM NEWTON

GREAT DISCOVERIES IN PHYSICS

Edward Speyer

Wiley Popular Science

John Wiley & Sons, Inc.

New York • Chichester • Brisbane • Toronto • Singapore

TO SYLVIA

This text is printed on acid-free paper.

Copyright © 1994 by John Wiley & Sons, Inc.

All rights reserved. Published simultaneously in Canada.

Reproduction or translation of any part of this work beyond that permitted by Section 107 or 108 of the 1976 United States Copyright Act without the permission of the copyright owner is unlawful. Requests for permission or further information should be addressed to the Permissions Department, John Wiley & Sons, Inc.

This publication is designed to provide accurate and authoritative information in regard to the subject matter covered. It is sold with the understanding that the publisher is not engaged in rendering professional services. If legal, accounting, medical, psychological, or any other expert assistance is required, the services of a competent professional person should be sought. ADAPTED FROM A DECLARATION OF PRINCIPLES OF A JOINT COMMITTEE OF THE AMERICAN BAR ASSOCIATION AND PUBLISHERS.

Library of Congress Cataloging-in-Publication Data:

Speyer, Edward.
    Six roads from Newton: great discoveries in physics/Edward Speyer.
      p.   cm.
    Includes index.
    ISBN 0-471-15964-6
    1. Physics. 2. Physics—History. I. Title. II. Title: 6 roads from Newton.
    QC21.2.S647   1994
    530—dc20   93–45342

Printed in the United States of America
10 9 8 7 6 5 4 3 2 1

Preface

We are in the midst of a scientific and philosophical revolution, on the scale of the Renaissance revolution against medievalism. This book gives an overview of the basic ideas of this revolution, arranged for the most part in historical sequence. They are exciting, fundamental, far-reaching, mind-stretching, and controversial ideas. Most people who live in revolutionary times are curious as to what is at stake; they want to partake of the excitement, at least vicariously. How can there be black holes, people in spaceships who age more slowly than we do, atomic dice that play by different laws of chance, observable events that do not have any specific causes, a basic graininess and fuzziness in physical quantities? What is happening to our traditional ideas of causality, reality, probability, scientific truth? Does the physical universe operate according to rules that make sense to us? Or should we give up trying to understand and picture, and be satisfied with predicting and calculating?

This book is partly an introduction to modern physics, partly a history, partly a philosophy of science. But mostly it is a road map in a land of ideas, guiding the reader through a newly discovered, exciting land. Since the reader does not speak the local languages, but would like to sample the culture and cuisine, it emphasizes the high points, simplifies the explanations, and avoids the technical jargon. Road maps show where things are, and how to get there. There are references for those who wish to dig deeper.

The first chapter, explaining the Great Clockwork view of the universe, especially as seen by Newton, is easily understandable to anyone who stayed awake through a high-school physics course; it includes a refresher (review) of Newton's laws. The next six chapters each describe a major road, the Six Roads from Newton. These six roads not only extend science beyond the Great Clockwork, but each indicates a new way of looking at the universe. Is the physical universe fundamentally composed of waves? Of fields? Of particles? Is it quantized? Is it ruled by strict causality or by chance? What does it mean if space, time, mass, and energy are all relative to the state of motion we are in? There is a little math, but nothing beyond high-school algebra.

Chapter 8 takes a quick look down some intriguing further roads, to give the reader a brief, necessarily incomplete, view of where physicists are working today. Chapter 9 presumes somewhat more technical background than the other chapters, and readers without combat experience (i.e., who have not worked on scientific or engineering problems) may find occasional patches of fog on the road. But not, I hope, frightening enough to make them want to turn back.

The tenth, and final, chapter resumes discussion of some of the Big Questions raised by ancient thinkers, who looked mostly to the stars. Newton moved the discussion significantly forward, but these questions are still being considered (with somewhat changed terminology) by scientists today. Final answers cannot be given, of course, but the discussion indicates the contributions that the physical sciences are making, and can make, in our searches. This involves looking beyond the laboratories and research institutes to how the scientific revolution is affecting our thinking and our world. The influence of science on our daily lives is great, but its influence on our thought, on the spirit of our times, is at least equally great.

The appendices deal with topics that were crowded out of the main text; readers who ''hate math'' may particularly enjoy Appendix D, which belittles the role of mathematics in modern physics. The heart of the revolution is in two theories, relativity and quantum, and it is the roads from Newton to these with which we are principally concerned.

Revolutions inevitably involve controversy; the reader is hereby notified that there are other interpretations than some of those given here, and that some of these other interpretations are held by the majority of physicists. Despite much publicity to the contrary, modern science has not developed an integrated world picture, nor even reached a consensus as to whether such a picture is possible or necessary. While the facts cited and the theories discussed in this book are standard fare, some of the interpretations are dissenting, minority views. For example, the omission of sophisticated mathematics is not only out of consideration for readers who ''can't stand math.'' Modern physics, in my opinion, tends to underrate the fundamental differences between mathematical operations and physical processes. Mathematical quantities are exact, and mathematical operations are either true or false; physical reality is not this neat. The chief sights described along each of the roads to and from Newton are experiments rather than equations, because these provide the hard facts on which interpretations of theory can be based, and reinterpretations discussed. I have posted little warnings at particularly controversial and/or heretical points, and the reader may enjoy comparing orthodox textbooks on these points. I run the risk, however, of being like the politician in a debate who was charged by his opponent: "He has told you many new and true things; but the things which are new are not true, and the things which are true are not new."

The paradoxes of quantum theory are still being debated, with some physicists disputing whether there even *is* a problem of interpretation. Einstein said, "God does not play dice with the universe," meaning that probability calculations, for example, dice odds, do not suffice for a complete physical theory. Niels Bohr, spokesman for the majority said, "Einstein, stop telling God what to do." Classical physics, stemming from Newton, makes sense, but it cannot predict or explain the results of many twentieth century experiments. Quantum theory predicts (calculates) these results, but does not explain them. Quantum theory does not make sense; it is acausal. When faced with two unacceptable alternatives, the best thing is to create a third. Chapter 9 explores how that might be done.

An ulterior motive of the book is to arouse the reader's curiosity, because curiosity is the key to learning and doing science. This book has the attitude: "It's fun to be fooled, but it's more fun to know." It thus carries a message of hope to those who want to believe that the physical universe makes sense. Not *should* make sense, but *does* make sense. The purpose of science is to demystify, not to leave people openmouthed.

In the scientific view of the physical universe culminating in, and growing out of, the work of Isaac Newton, everything makes sense; if physical phenomena do not obey the laws perfectly, that is because we scientists do not measure, or calculate, the phenomena perfectly. Man is the imperfection in a perfect universe, but his Reason, in the traditions of Aristotle, Spinoza, Diderot, Newton, Robespierre, Dewey, and Einstein (to name only a few) is adequate.

However, since Newton's time, a number of great cracks have been found in the Great Clockwork, indicating that important aspects were left out of account. Six of these cracks, or roads along which physical science has developed away from the Newtonian viewpoint, are examined in this book. The theme of the book was expressed by Einstein: "God is subtle, but not malicious." Einstein expressed himself in religious terms for emphasis; he meant that the whole physical universe, including the quantum domain, can be understood if we are willing to explore its subtleties; there are no dirty tricks, contradictions, or illogical stunts in nature. This doctrine might be called "atheistic monotheism" because, while avoiding appeal to a First Mover, Great Planner, Supreme Purpose, or Father in Heaven, it nevertheless affirms a fundamental belief in unity and order and intelligibility. Physical phenomena are not segregated into two exclusive categories: caused and uncaused, nor objective and subjective. Observable events are always composed of (due to) comprehensible (even if not measurable) processes. The universe is orderly, and its truths, although not simple, are comprehensible to the human mind, the inquiring human mind. The knowability of the physical universe is an unproven assumption, a matter of faith, but without it physics is reduced very nearly to the level of a "trip" in a special, university-approved, drug culture.

Contents

1

The Newtonian Universe

Early Roads

In the beginning, as the saying goes, science and religion were indistinguishable. The fearful unknowns of the environment were imagined in manlike and animallike forms, so that the strange became identified with the familiar. This is the essence of explanation, whether scientific or religious. Anthropomorphism is a simple type of theorizing: rain is caused by rain gods; thunder by gods of thunder; storms, volcano eruptions, and earthquakes by gods' anger; sickness by evil spirits. Unknown aspects of the world are associated with known aspects; human events are manipulated by gods who are not very unlike the humans who created them.

What we today would call scientific notions developed within religion, not in opposition to it. The ancient scientists, including the Greeks, and the early modern scientists, were religious, and some of them wrote on religious subjects. A vestige of anthropomorphism was the explanation of why air rushes into a low-pressure container: "Nature abhors a vacuum." We can borrow this formulation for modern science: Nature abhors perfection. (There is always noise, uncertainty, and limited resolution.)[1]

---

[1]In a later section of this chapter, "Laplace's Demon" is exorcised.

The sciences tended to develop with decreasing distance between the areas of search and the questions being researched. Astronomy's area of search is the stars, zillions of miles away from the scene of the Big Questions being asked: Who are we? Where did we come from and where are we going? What is our purpose and place here? What does the future portend? Imagine a detective called to a murder scene, who walks into the room and sees the body on the floor, then leaves and takes the first plane to Tibet. Why go so far away to search for clues to the questions arising here at hand?

Gradually, through the centuries, scientists extended their research into areas closer to home: biology, chemistry, geology, then sociology, anthropology, and psychology. Only gradually did we become curiously aware of ourselves. We wondered about the stars, the animals and the volcanoes, but we were reluctant to ask "why?" about our own feelings.

Is man at the center of the universe? The ancients thought so, and put the earth at the center of the solar system and of the heavens, a geocentric view. This view reflects the human desire, which has inspired science through the ages, to make the universe comprehensible. Whether or not we are seated at the center, however, it is still our show, our universe. The questions, the moral values, the observations, and the interpretations are ours. The geocentric view put humans on center stage because the Creator would have it no other way. Today we might express this idea as: Man is the measure of all things.

In ancient times, astronomy was chiefly astrology. The astrologers are still in business, still using the same methods; they have learned very little in 2,500 years. With the ancient Greeks, however, scientific thinking entered an important new stage. The Greeks believed not in the mysteriousness of the universe but in its understandability, that it makes sense when we interpret it properly. There were, of course, many Greek thinkers; the most "modern" was Archimedes (287–212 B.C.), whose deduction of the physical principle that bears his name is a model of scientific reasoning worth studying today. Mention should be made also of Democritus (460–370 B.C.), who developed an atomistic theory of matter. Eratosthenes (276–194 B.C.) calculated the earth's circumference by comparing the angular elevations of the sun (as shown by shadow lengths) at different latitudes on the same day, and he got a result far closer to ours than did Columbus or Magellan. In fact, if Columbus had had as good an estimate of the earth as Eratosthenes had, he would never have mistaken his landfalls for India. Aristarchus (300–250 B.C.) developed a heliocentric model of the solar system. Ptolemy (70–147 A.D.) compiled, and added to, the ancient knowledge of geocentric astronomy. And there were many more.

From Aristotle to Galileo

Perhaps the greatest thinker and systematizer of ancient Greece was Aristotle (384–322 B.C.), whose writings dominated the thought of the Renaissance more than 1800 years later. Aristotle was the real intellectual antagonist with whose thought-experiments Galileo and Newton had to contend.

Undoubtedly, Aristotle was one of the great geniuses of all time. One of the characteristics of a genius is that his chosen field is so imbued with his work that anyone wishing to transcend that work, to carry it into new areas, is practically forced to deal with the work of the genius. That is to say, we can well begin by arguing with the genius. In science, the quality of our work is closely related to the quality of those thinkers we disagree with.

With the wisdom of 2,000 years' hindsight, we can easily see that Aristotle made many errors. Nonetheless, some of our current wisdom can be clarified by starting with Aristotle's writings on a given subject, and then pointing out how scientific knowledge has evolved from them. Lesser people's contributions are often ignored and forgotten; geniuses have to be taken account of, because right or wrong, they address central issues.

Aristotle saw an inherent tendency of four basic substances or essences— earth, water, air, and fire—to assume their natural and proper place. Earth's place was at the bottom; water rested on the earth; air on top of that, and flames rose toward the stars, which were of fire. Stones, being earth, fell to earth, and sank through water. Air bubbles rose through water. Each of these essences naturally moved toward its proper place, as though nature itself was a causative factor, without any other cause being required. Between the stars was a fifth essence (which was later to be called the ether); we get the word "quintessence" from "fifth essence."

Among the many subjects he addressed, Aristotle wrote about moving bodies. He apparently thought that motion (velocity) was due to force. If an arrow flew through the air, he reasoned, a force was pushing it, and the faster the arrow flew, the greater was the force propelling it. When the force stopped pushing it, it stopped moving. Aristotle reasoned that if it weren't for air friction, heavy bodies would fall at the same speed as light bodies. He concluded that a vacuum was impossible! Aristotle's disciples used this as giving Aristotle's authority to the thesis that heavy bodies fall faster than light ones.

Let us look at how Aristotle's concept of force, which is partly correct and partly incorrect, explains the interactions between the five essences. A force (gravity) pulls a stone down through the water, but what force pushes bubbles up to the surface? What force do bubbles have, or a piece of wood that has been submerged and then released? When we dive into a swimming pool, what force brings us up to the surface? And how about a hot air balloon, or one filled with hydrogen? Aristotle's conception of force does not adequately explain the motion in these cases, but it was a fundamental milestone down the early roads that led to Newton's physics.

Perhaps Aristotle's most lasting contribution was his logic. He worked out the rules of logical syllogisms, and much of what he taught is still taught today. His logic is two-valued: a thing is either A or not-A; a proposition is either true or not true. If we assume something is true, and then show that that leads to a contradiction, then we have proved the contrary of our assumption. This is called indirect proof, and is much used in mathematics. Other applications include computers, which use a binary (two-valued) code based on Aristotle's logic. A digit is either 0 or 1; a circuit is either open or closed; the IBM card is either punched or not punched. According to this logic, questions are answered by either "yes" or "no."

Two-valued logic has come under question in modern times, however, and has been ridiculed: "That which is, is; that which is not, is not; and anything else is the work of the devil." Or, as the song has it, "Don't mess with Mr. In Between." Indeed, scientists must mess with Mr. In Between. Who would suspect, for example, that between a state of matter in which molecules are bound together (a solid), and a state of matter in which they are not bound (a gas), lies another stable state, a liquid?

Two-valued logic obscures our ability to see that a whole can be quite different from the sum of its parts. Also, a large change in the magnitude of a physical quantity can change its fundamental character. In dialectical terms, there can be a transformation of quantity into quality, such as when an increase in explosive nuclear fuel exceeds the critical mass, or a light blow (e.g., tap on the head) becomes forceful enough to be lethal. As important as Aristotelian logic remains in contemporary science, we will see that we need a more subtle logic with flexible categories as we travel the roads from Newton. Existence, truth, and causality need not be airtight categories.

If it were not for the Arab scholars, western civilization would have lost even more of the ancient writings than it did during the Middle Ages. History does not progress continuously; societies and intellectual disciplines can retrogress. However, around 1320, William of Occam wrote that we ought not to admit into the explanation of any phenomenon any more unproven assumptions than are strictly necessary; we are to shave off any excess assumptions; hence this parsimony principle is known as Occam's razor. William may have had a beard down to his knees for all we know, but we still shave off extraneous postulates and hypotheses in his name. Einstein's theory of relativity is scientifically simpler than Newton's (although the mathematics is more complicated) because Einstein dispenses with (shaves off) Newton's assumptions of absolute motion and absolute time, as we shall see on the Relativity Road.

A clear example of Occam's razor in action is the replacement of the Ptolemaic model of the solar system with that of Copernicus, and later, of Kepler, Galileo, and Newton. Ptolemy, following Aristotle, placed the earth at the center (geocentric model), and arranged the orbits of moon, sun, planets, and stars around it (the heavenly spheres). Today, we believe that the universe is expanding, with the more distant galaxies receding from us at high velocities. Does that imply that we are at the center, i.e., at the site of the Big Bang? Why, out of all the places in the universe that it might have happened, did the Big Bang occur *here*? The odds against our getting front-row center seats by happenstance are stupendous, and indeed unacceptable. However, we don't *have* to embrace the Ptolemaic model in order to believe that the universe is expanding. The spots on a polka-dotted balloon, as it is inflated, all move further apart, no matter which spot is "the earth." Similarly, the raisins in a cake, when it is rising, all move apart from each other, without regard for which is our "home raisin." The expanding universe is conceived as looking pretty much the same from most of the galaxies, not just ours. Today's scientists would reject a theory or model which gave us a preferred position without experimental justification.

One of the big problems with the Ptolemaic model was the retrograde (backward) motion of the planets. Every so often a planet reverses its direction among the stars. This is not sudden, and it happens according to regular periods, but the planet's movement among the stars, night after night, slows down, and then reverses for quite a few nights, before resuming its normal progress. Mars is the biggest offender. Ptolemy placed each planet on a smaller circle (called an epicycle), which rotated relative to the heavenly sphere for that planet. Thus, the planets were on circles rolling relative to other circles. These epicycles certainly disregard Occam's razor. To justify his belief that the earth is at the center of everything, Ptolemy had for evidence that we do not feel any motion on the earth, but for the epicycles, Ptolemy was on weaker ground.

Our current understanding owes much to the work of Copernicus (1473–1543), who developed a heliocentric model of the solar system. For more than a century, the followers of Ptolemy and the followers of Copernicus argued the merits of their two models. In every scientific argument—and much of our progress has been made through arguments—the resolution turns on a few issues, or decisive experimental facts. There were two such issues between the geocentrists and the heliocentrists. One was the retrograde motion of planets, referred to above. Copernicus explained the backward motion of Mars as due to the earth overtaking Mars, since earth's orbital period (one year) is shorter than that of Mars. Mars appears to move backward the way a car appears to move backward to another car passing it on the racetrack, a matter of relative motion.

The second decisive issue also involves relative motion; this is parallax, which is apparent motion between near and far objects, as seen by a moving observer. For example, when we drive along the highway, the moon moves through the trees, and when we move our head in the theater, the head of the person in front of us moves across the stage. How is it, asked the geocentrists, if the earth is moving, that we do not observe motion of the nearer stars relative to the more distant stars? The heliocentrists had no good answer to this question until the nineteenth century, when the tiny parallax motion of a few near stars was, with difficulty, measured. The only answer the Copernicans could give before then was that the stars are so very far away that parallax is too small to be observed, which turned out to be the case. It was, however, considered a weak answer at the time.

In dealing with the retrograde motions, Ptolemy had an advantage over Copernicus, in that since he did not "shave" with Occam's razor, there were more ways to tailor (revise) his model to fit the observational data. In later versions, the Ptolemaic model even had epicycles within other epicycles. By ad hoc adjustment of distances and velocities the geocentrists were able to match the accuracy of the Copernican predictions, which did not correspond perfectly to the observations either. The problem for the Copernicans, however, turned out to be that Copernicus, like Ptolemy, had insisted on "perfect" orbits, that is, perfect circles. It was not until Kepler (1571–1630) that anyone perceived that more accurate predictions of planetary positions required *elliptical* orbits.[2] So it isn't always the better theory that wins the argument at first.

The greatest champions of the Copernican system were Kepler (of whom more in Chapter 10) and Galileo Galilei (1564–1642). Galileo's contributions cover a wide area. Best known, perhaps, are the discoveries he made using telescopes. He did not invent the telescope, but he was the first to use it for scientific work. He made his own telescopes, and observed the four biggest moons of Jupiter, still known as the Galilean moons. These moons caused great excitement, because they revolved around Jupiter, showing that the center of all celestial revolutions did not have to be the earth. They also provided partial answer to the Ptolemaic question (argument) as to why, if the earth moved around the sun, the moon followed along. A fuller answer had to await Newton's law of gravity. Galileo (and Kepler) never quite grasped the concept of gravity.

Galileo also observed the phases of Venus, which resemble those of the moon, and have a similar cause, varying angles of illumination by the sun, as seen from the earth. He observed the craters on the moon; a long-lasting argument arose as to whether the craters were volcanic or due to meteor impacts. This argument ended in a draw (both types of craters seem to exist) when American astronauts brought back samples of moon rocks.

---

[2]Kepler's work is summarized in Chap. 10.

Among Galileo's other astronomical discoveries were the existence of sunspots, which appeared to Galileo's critics, of which he had many, to be imperfections, or blemishes, on the Creator's work. (Photographs of the sun showing sunspots look like a towel after a small boy has dried his dirty hands on it.) Galileo's critics were by no means limited to officials of the Church. "That thing," it was said, pointing to the world's first practical telescope, "is a thing of untruth. It makes things appear as they are not." The critics maintained that no man of virtue would look through it, and certainly wouldn't believe anything seen through it.

It should not be concluded from this, or from Galileo's experiments with falling bodies, that science diverged from religion principally because scientists were fond of experimentation. Actually, the early scientists, including Galileo, diverged in a rationalist direction, urging that things could be explained rationally, without recourse to mysticism, miracles, or authoritative writings. Experiments were performed, not in opposition to theorizing, but in support of it. In those days, the answers to serious questions were generally sought in the Bible, in Aristotle, and in Church writings. If an answer couldn't be found there, then you probably shouldn't be asking the question. Galileo wasn't afraid to say, "I don't know" with the implication, "but we'll investigate and try to find out." When orthodox thinkers said "I don't know," they usually implied, "and we don't want to." Galileo's troubles with the Inquisition were attributed to his Copernican teachings, although several cardinals accepted them, and Copernicus himself had been a churchman. However, one suspects that the root trouble was Galileo's attitude that Truth could be worked out by an individual thinker and observer. This dangerous attitude was made worse by Galileo's publishing in Italian, instead of the customary Latin, so that common people, who read only Italian, might have their faith disturbed.

Galileo explained many of his scientific ideas in the form of dialogues between three characters: Salviati, who is obviously Galileo himself, explaining everything; Sagredo, the intelligent layman who asks the right questions and is duly convinced by Salviati; and Simplicio, who comes off the worse in each argument. Simplicio, however, is not an ordinary, Dr. Watson type of fool; he is, in fact, a learned Aristotelian, doing his best to uphold Aristotelian doctrines.

Galileo saw through Aristotle's major mistake in dynamics, and grasped the truth: an arrow flies through the air at constant speed as long as *no* force is acting on it. Air friction, which is a retarding force, is what slows it down; if there were no air friction, the arrow would keep on going forever (except for gravity, which is also a force). A hockey puck would slide forever if there were no friction with the ice. The velocity of the arrow builds up from the force exerted by the bowstring when it is shot. Force thus correlates not with velocity, but with *change* of velocity.

Galileo did not put this idea quite as sharply, nor in mathematical form; that remained for Newton. But Galileo pierced Aristotle's armor, and prepared the way for Newton's first two laws. The only reason that a feather falls slower than a stone is because of the greater effect of air resistance on the feather. On the moon, which has no air, or in an evacuated tube (a cheaper way to do the experiment than sending an astronaut), a feather and a stone can be seen to fall together.

At one point in one of Galileo's dialogues, Salviati uses the following argument: If a heavy body were to fall faster than a light one, what would happen if a heavy and a light body are fastened together, so that they fall as a single body? If the combined body falls faster than either falls alone, how can the speedup be explained as due solely to our fastening the two masses together? And if it falls at an intermediate (weighted average) speed, then it is falling slower than its heavier component fell by itself. This would mean that a heavier (composite) body falls slower than a body that weighs less. Salviati concludes that all bodies must fall at the same speed, regardless of weight (mass). The perspicacious reader will spot the weakness in this argument. The conclusion to the argument is correct, but we will reach it along a better road with the aid of Newton's second law and law of gravitation.

Newton's Laws

The climate of scientific inquiry in Italy was chilled by the attitude of the Church; the recantations forced on Galileo were more than symbolic. Original thinking found a warmer climate in England. Modern science got its biggest start as the hobby of upper-class Englishmen and intellectuals, who had not only leisure, money, and education, but also freedom from the Inquisition. The Royal Society, as they sensibly called their organization (with the king as honorary member, of course) had many talented members, but head and shoulders above all the rest was Isaac Newton (1642–1727).

Newton's great masterpiece, *The Mathematical Principles of Natural Philosophy*, was published in 1687, a date which has something like the significance in physics as 1776 has in American history. *Principia*, as it is most commonly called, has probably had a greater influence on our society than any other book except the Bible. If a genius is a person who sees things in ways that his contemporaries cannot yet see them, but will see later, then Newton certainly ranks with Aristotle, Da Vinci, Shakespeare, and Mozart. There were those who considered Newton to be the happiest man who would ever live, because it could be given to only one man to discover the laws of the universe. Indeed, Newton's achievements are such that the English poet Alexander Pope wrote: "Nature and Nature's laws lay hid in night; God said, Let Newton be! and all was light."

Newton invented the calculus, which he needed for computing planetary orbits and other problems, in addition to other mathematical discoveries. Some of his work was in optics, which we will encounter on the Wave Road. But the heart of Newton's system of moving bodies (dynamics) and the base for much that followed in physical mechanics, is in his three laws of motion, plus his law of universal gravitation. The three laws, translated here freely from the Latin, state:

First law: the law of inertia
A body at rest, or in uniform motion in a straight line, will so continue unless acted upon by an outside force.

Second law: the law of acceleration
If a body is acted upon by an outside force, it will accelerate in the direction of that force, and the acceleration will be proportional to that force, and inversely proportional to the mass of the body. That is, force equals mass times acceleration: $F = ma$.

Third law: the law of recoil
When a body exerts a force on a second body, a second force, equal in magnitude and opposite in direction, is exerted by the second body on the first body. Action is equal to reaction.

The first law says that a moving body keeps on moving, and a body at rest keeps on resting. What changes this situation is a force, a push or pull of some kind. A moving automobile has to be braked in order to slow it down, and it has to be given a thrust, from the engine, in order to start it moving or to speed it up. If there is no retarding force, such as friction, the body, like a spacecraft coasting in deep space, will keep on going forever. Newton drew the distinction, not between motion and nonmotion, as the ancients had done, but between uniform motion, including nonmotion (rest), and *accelerated* motion. We therefore have to distinguish carefully between velocity, which we measure in meters per second, and acceleration, which is change of velocity, measured in meters per second per second (m/sec^2). When we accelerate, we gain so many meters per second, in each second.

Velocity and acceleration each possess two properties, which are measured separately: magnitude and direction. If a body is moving at a constant number of meters per second, but is in orbit around another body (as the earth is), then it does not have a uniform velocity, because its direction is changing over time. Going around curves, being in orbit, or spinning (rotating), are all considered accelerated motions, just as picking up speed and braking are.

Newton's first law treats a body at rest and a body with a uniform velocity as equivalent. This means that we cannot distinguish between them on an absolute basis, but only relatively. Whether you are at rest or in uniform velocity makes no difference according to the physical laws, and it is arbitrary (up to you) which you want to consider yourself. This is the principle of relativity, applied to uniform motion; it is sometimes called Newtonian relativity. Einstein extended this principle in several ways, as we shall see on the Relativity Roads.

As illustration of Newtonian relativity, consider a baseball team en route to a game aboard a train, boat, or airplane. The pitcher wants to practice his curve and drop balls, and tells the catcher to stand at the other end of the aisle. Does it make any difference in the pitched balls if the vehicle is moving or standing still? Not if the velocity is uniform (a smooth-riding vehicle). Can the pitcher even *tell* whether he is at rest or not (with respect to the earth) without looking out the window? No.

Here is a little test question. A sailor on top of the mast drops a tool. The boat is moving with uniform velocity. Will the tool hit the deck at the foot of the mast, or will it hit nearer the stern because the boat moves while the tool is falling through the air? Remember, the tool has the same forward velocity as the boat, before it is dropped, and it keeps that velocity while it is falling because it has inertia. So the answer is: at the foot of the mast. Converses of the first law are also valid:

- If a body is at rest or in uniform motion, the forces acting on it must be zero (balance out), and the body has no net external force acting on it.
- If a body is acted upon by a force, it will change its state of rest or of uniform motion. This can be regarded as a qualitative definition of a force.

With a little stretching, the idea of inertia can be extended. Living organisms seek to perpetuate themselves, by self-preservation and by propagation of the species; in this view, mutations, which are usually counterproductive, are analogous to noise and random fluctuations in physical processes. All physical processes have some inertia because they have finite response times. Unchanging physical systems, and physical systems undergoing change, tend to so continue until some new causal factor takes effect (kicks in). This idea of universal inertia can be applied to graphs of physical processes, such as voltage against resistance, or temperature against pressure. We can conjecture that whenever the slope of the graph of a physical process changes its direction, some other factor may have become significant. Conversely, uniform change is represented by a straight line graph, and no change is represented by a horizontal graph. Every time an experimental curve changes direction, we can look for a new causal factor, or else an old factor recurring (as in periodic phenomena).

Newton's second law makes the first law quantitative. Forces are measured, not by velocities themselves (as Aristotle thought) but by the accelerations, or changes in velocities, they cause. If a truck and a Volkswagen
each run out of gasoline on a level stretch of road, we have to push harder
(greater force) to get the truck moving, that is, accelerate it from rest, than
to get the Volkswagen moving to the same velocity. This is because the
truck has greater mass and inertia, and the change in velocity is inversely
proportional to mass, according to Newton's second law.

Does this reasoning apply to *falling* bodies? You bet! But some might
see a problem, for if acceleration is proportional to the force causing it, as
the second law states, and the earth pulls *harder*, by its gravity, on a heavy
body than on a light one, then how come the heavy body does not fall *faster*
than the light one? Are we back in the swamp of Aristotelianism? No way.
A heavy body *needs* more force in order for it to be accelerated than does
a light body, just as with the truck and the Volkswagen. The second law lets
us escape from Aristotelianism because the proportionality constant between
a force and the acceleration it causes is the mass (inertia) of the body being
accelerated. This revolutionary idea will become clearer after we have Newton's law of gravitational force to help us.

First, however, let's finish with the laws of motion, and turn to Newton's
third law, which says that forces always occur in pairs, equal and opposite.
The book resting on the table pushes down on the table, and the table pushes
up on the book. Notice that each force acts on a different body—one on the
book, the other on the table—so that they do not cancel each other. If the
book pushed harder than the table, either the book would sink into the table
surface, or the table would collapse; if the table pushed harder on the book,
the book would fly into the air. So the two forces are equal, symmetrical.
The same reasoning applies to me when I stand on the floor; if I push down
harder than the floor pushes back, the floor is giving way. Similarly with
bodies that do not touch, but push or pull each other. The earth pulls on the
moon, and the moon pulls on the earth. The pulls are equal in magnitude,
but not in effectiveness, because the moon is much less massive than the
earth, and so the pull affects the moon more, and it orbits the earth. The
third law can also be illustrated by the recoil of a rifle, and it lets us see
how rockets can be accelerated in space; the rocket pushes against its own
exhaust gases.

Consider the following more complicated problem. Suppose there are two
similar small boats equidistant from a pier. In each boat is a man who wants
to bring his boat in. One man is pulling on a rope, the other end of which
is tied to a stanchion on the pier. The second man is pulling on a similar
rope, but the other end of his rope is held by a friend, standing on the pier.
All three men pull with equal force. How much faster does the second boat
come in than the first? They come in at the same speed, and arrive at the

same time, because the stanchion pulls too, and with equal force. The two friends each pull in only half as much rope as does the man working alone, because the friends are pulling the rope out of each other's hands. Consider what would happen if the stanchion did *not* pull, for example, suppose it breaks off and is pulled into the water. Or suppose the friend on the pier is not anchored firmly like the stanchion, but slips on a dead fish and is pulled into the water; in that case, the stanchion pulls and the friend does not.

Here is another third law problem. A monkey is hanging on a weightless rope, which runs through a frictionless pulley. On the other end of the rope is the monkey's mother-in-law, who weighs the same as he does. The monkey wants to escape his mother-in-law. Should he loosen his hold on the rope, and let her drop? No, because then he would drop also, and at the same speed. Should he climb up the rope to the pulley? No, because in pulling so that he can climb, he will lift his mother-in-law too. The answer is that nothing will help him get away from her.

The Law of Gravity

In Chapter 10 the solar system is described briefly, and there we give Kepler's three laws of planetary orbits (see Figure 18). Kepler's discovery of his three laws preceded Newton, and Newton used them in finding his three laws of motion. But the force that held the planets to the sun was missing, and Newton supplied that. Newton's law of universal gravitation is

$$F = G\,\frac{m_1 m_2}{d^2},$$

where F is the force of attraction (in each direction, as required by the third law). G is a universal constant, having the same value everywhere. The two masses m_1 and m_2 are those that are attracting each other, say the earth and the moon, or two glasses of water. The distance d between the two masses, or more precisely, between their centers of gravity, is discussed below.

Every body, or particle, or quantity of mass, in the universe attracts every other body in the universe with a force proportional to the product of their masses, and inversely proportional to the square of the distance between them. The quantity of mass can be as big as a star or a galaxy, or as small as a molecule or an atom. It can be liquid, gas, or solid, alive or dead, hot or cold. The food on my plate attracts my hand, but the pull is too small for me to feel. As the distance between two bodies increases, the force of attraction between them decreases, not linearly, but as the second power of the distance. Thus, if two bodies that were 1 meter apart are moved to 2 meters apart, they attract each other only one-fourth as much. The same kind of inverse square law applies to light; if we move a candle twice as far away from the book we are reading, we get only one-fourth the illumination on the page. Inverse square laws apply to magnetic forces, to electric forces, to sound intensities, and to the amount of heat we get from a small hot source. What is the reason for this similarity?

Because in each case we are really concerned with *area*, the area of an imaginary sphere whose center is the source, and whose surface includes our detector. (We are not talking of the intensities felt by the eye or the ear; our sense organs are highly nonlinear.) The area of a sphere increases with the square of its radius; that's where the square law comes in. Notice that we are using a geometric property of space to calculate a physical relationship. Notice, too, that the "stuff" emitted by the source goes out radially, taking the shortest distance; physical processes tend to follow the easiest, shortest, fastest routes. Such routes are called geodesics; principles of geodesy are found throughout physics, for example, least time, least action, least constraint. We shall meet some of them later.

G, the universal gravitational constant, (not to be confused with g, the local acceleration due to earth's gravity) never changes. Its numerical value depends only on the units we use, not where we measure it; it is the same on the earth, near the sun, and behind the moon. Whether or not G represents a characteristic of the universe (as c, the velocity of light, seems to) is an open question.

When two masses of significant size attract each other, we have a complication. Consider the earth and the moon. A piece of the earth facing the moon is nearer the moon than a piece of the earth on the far side. Also, different parts of the moon are at different distances from any single part of the earth. The strength of the gravitational attraction is different for each pair of masses we consider, if their separation is different. Since we can divide the earth and the moon into "skillions" of little pieces, we have a double infinity of calculations to add up in calculating the total gravitational attraction.

Newton saw this problem, and solved it for two perfect, uniform spheres. For that nice case, the symmetry of the spheres is such that all the greater-than-average attractions are balanced exactly by less-than-average attractions. We end up having only to consider that each sphere has all its mass concentrated at its center; that gives us only the distance between centers in calculating the gravitational force between them. Using uniform spheres is thus obligatory if we want to avoid a very messy calculation, probably requiring a computer. An accurate calculation of the attraction between two boxes, or between a football and the pass receiver, is a horrible mathematical problem.

Now that we understand something about gravity, we can see, and show, why heavy bodies and light bodies fall at the same speed, as Galileo said. From Newton's second law we have

$$F = ma,$$

where m is the mass of the falling body. We can also write the force of the earth's attraction on that body, using Newton's law of gravitation:

$$F = G \frac{mm_E}{R^2},$$

where m_E is the mass of the earth, and R is the radius of the earth, since that is how far we are from the earth's center of gravity. We now have two equations for the same force, so we set them equal to each other. The mass of the falling body cancels out, which leaves us with

$$a = \frac{Gm_E}{R^2}.$$

This shows that the acceleration of any body falling near the surface of the earth (neglecting air friction and the effect of the earth's spinning) is given by three *constants*: G, m_E, and R. Galileo was therefore right; the mass of the falling body makes no difference. These three quantities also show that a heavy and a light body will fall together near the surface of the moon, but at a slower speed than they do near the earth because the attracting mass and the distance from the center of gravity are different on the moon. On Jupiter, heavy and light bodies will fall together, but faster than on earth or on the moon. The much greater mass of Jupiter more than compensates for its greater radius (distance from the center of gravity).

On earth, we give the acceleration of a falling body a special letter, g. In most places on earth (a little less on mountaintops, and near the equator) this equals 9.8 meters per second per second. We write this $g = 9.8$ m/s^2. It means that each second that a body falls, neglecting air friction, it falls 9.8 meters per second faster, and 9.8 meters further, than it fell during the preceding second.

Newton applied his law of gravity to the earth-moon system, and solved the riddle of the tides. One aspect of the riddle is that there are generally *two* high tides at each place on the earth every day, but the earth rotates only once on its axis in that time. A second puzzling aspect is that if we substitute the mass of the sun and its distance from the earth into the law of gravity, and then repeat the calculation using the mass and distance of the moon, we find that the center of the sun pulls on the center of the earth 178 times more strongly than does the center of the moon. This makes sense, because the earth is twirled around the sun, while the moon is twirled around the earth. But then why do the tides more strongly follow the moon?

Newton found that the tides are not due to a direct pull on the earth, but to the *difference* in pull on different parts of the earth, a second-order effect. If the moon pulled equally hard on all parts of the earth's surface, there would be no tides. The tides are due to the near side of the earth being pulled harder by the moon than the far side of the earth. The earth is so much nearer to the moon than to the sun that the moon's gravitational field on the earth has more variation than the sun's gravitational field does. There are tidal effects on the side of the earth nearest the moon and also on the side of the earth farthest from the moon, so we get two tidal cycles a day.

This does not explain, however, why the tides run as high as 34 feet in the Bay of Fundy, but only about 10% of that farther down the Atlantic coast, or on the Pacific coast. The Bay of Fundy has a resonance; the time it takes a wave of water to travel the length of that bay is a harmonic of the earth's rotational period with respect to the moon. If we were willing to spend the money, we might be able to tune (i.e., dredge and fill in) a bay, like the Chesapeake, to have higher (or lower) tides than it now has. Everyone has heard of the soprano who can break a wine glass by singing one of its resonance frequencies at it. We shall encounter the subject of resonance, again, on the Wave Road.

Newton's Laws for Astronauts

To get a feel for Newton's laws, let us apply all four of them to an astronaut orbiting the earth in a space capsule. The newspapers try to help the public, which has not had your advantage of reading this book, by explaining that the astronaut does not fall downward and hit the earth because the force of gravity, which is pulling him down, is exactly balanced by the centrifugal force tending to make him fly outward from the earth. The canceling of these two factors also "explains" the astronaut's weightlessness. This is piffle; Newton is entitled to equal time, for rebuttal.

Newton's third law says that if the earth pulls *down* on the astronaut and the capsule, then they must be pulling *up* on the earth. Those are the equal and opposite forces; their magnitude is calculated directly from the law of gravity. The capsule shows the effects (acceleration) due to that force much more than does the earth. The acceleration of an object is inversely proportional to its mass, according to the second law, and the mass of the earth is tremendously greater than that of the capsule.

Now let us look at how Newton's *first law* applies. Which way does the astronaut tend to fly away from the earth; tangentially or radially? When

David whirled his sling before letting the stone fly at Goliath, he needed to know whether the stone would fly tangentially or radially when the force of the thong pulling the stone inward (corresponding to the gravity pull on the orbiting astronaut) was released. The answer is tangentially, because the stone continues in the same direction as it was going at the instant it was freed from the force of the thong. In other words, David's stone and the orbiting astronaut have inertia, and tend to fly in a straight line. But if the astronaut tends to fly tangentially, gravity cannot be balancing his weight, because the force of gravity pulling him downward is *perpendicular* to the tangential direction. When one force is perpendicular to another, the two forces cannot cancel. So what force is canceling out gravity?

There is no such force. Gravity is pulling on the astronaut all the time. That's why he is in orbit; he is *falling* over the horizon, around the earth. He has a horizontal velocity component, given him by the rocket's second stage, which put him into orbit after he got up to a satisfactory distance above the earth. But this horizontal velocity component is combined with the motion produced by gravity. The astronaut's orbit is the continuous combination of two velocities; the one from the rocket second stage, and that produced by gravity, which in ordinary cases we call falling. The continually changing direction of the space capsule is an acceleration, which must be due to a force (Newton's second law again), and that force is earth's gravity. So the astronaut is falling even though he gets no closer to the surface of the earth. If the astronaut were not falling, that is, if gravity were somehow "turned off," he would fly off tangentially.

Let's look at why free-fall involves weightlessness. Notice that we are not talking about a disappearance of mass, but of the disappearance of a force to hold up a mass against gravity. Suppose I get into an elevator near the top floor of the building, holding a heavy shot in my hand; if I let go the shot, it will fall on my foot. It is not weightless. Now suppose the elevator cable breaks, and I am in free-fall. So is the shot. If I let go the shot, it will not hit my foot, because we are all, regardless of our different masses, falling at the same speed. The shot will stay right were it was, falling with me. I do not exert any upward force on it to keep it from hitting my foot. It is weightless. What more convincing way to show that the shot is weightless than that it stays in place without being held up? If I am standing on a scale in the elevator when the cable breaks, the scale reading will go to zero, because the scale is falling at the same rate that I am, and so does not push upward against my feet. So I am weightless, too. What will happen to the shot, scale, and me when we reach the bottom and deaccelerate is, of course, a different and sadder story.

But elevators move only vertically. To relate what happened in the elevator to the orbiting astronaut, we have to look at something that also has a horizontal velocity component like a cannonball fired horizontally off a cliff overlooking the sea. The cannonball will splash down at the same instant that it would have splashed if we had simply dropped it over the cliff. It will splash down further out from shore, of course, but its vertical fall rate is the same whether it has a horizontal velocity component or not. Now let us repeat the experiment, but this time fire the cannonball using a *great deal* of gunpowder, so that it has enough range to encircle the earth. The cannonball keeps falling, as before, but the sea falls away beneath it, because the earth is spherical, not flat. The cannonball is *in orbit* because it is falling around the earth, because of earth's gravitational pull on it, combined with the horizontal velocity it was given by the gunpowder. The same holds true of the orbiting astronaut.

How much gunpowder is needed? Suppose we use too much gunpowder, so that the cannonball has a still greater horizontal velocity out of the cannon. It will fall at the same rate, g, as before, but it will not follow the curvature of the earth. Instead, it will fly off at an angle, never to return. The rocket forces on a space capsule have to be calculated and controlled very carefully, so that the orbit is an ellipse, not a parabola or a hyperbola. An ellipse is a closed curve, which keeps going around and around, but a spacecraft in a hyperbolic orbit will never come back to earth.

Figure 1 is based on a drawing of Newton's[3]; it shows different possible orbits from a launching from a mountaintop. The situation is very like the one we discussed, of firing a cannonball off a cliff over the sea. Newton's view of planetary orbits, and of the universe as a Great Clockwork, spread over Europe and the world. *Principia* is written in Latin and is not easy going even in English translation; it was, however, widely popularized, much as the difficult relativity theory of Einstein was, more than two centuries later. Newton's theory was even published in a ladies' edition. Most people may not be interested in physics, but they are interested in the Big Questions concerning human existence, and maybe purpose, in the universe, and they will read material that seems to give even partial answers.

In the astronaut discussion, no mention was made of centrifugal force. What happens to centrifugal force is interesting, and will help us understand the General Relativity Road. If you are in a car going around a curve too fast, you are thrown outward against the door. If the door suddenly opens, do you fly out tangentially or radially? If a policeman is watching, he sees you fly tangentially because, whether you obey the traffic laws or not, you still obey Newton's first law. But you see yourself flying away from the car, radially. This is because you continue to move in the direction the car was moving at the instant you flew out. You are chiefly concerned with your velocity relative to the car, which is continuing around the curve, while you are going in a straight line. In other words, in the reference frame of the

---

[3]De mundi systematic, 1731 edition.

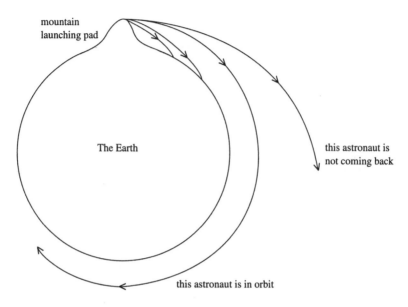

Figure 1 Plan for orbiting satellite (based on Newton's sketch).

car, you fly out radially; in the reference frame of the earth, and the policeman, you fly out tangentially. The observed direction of motion is relative
to the motion of the observer. This idea will be extended on the Relativity
Roads.

Laplace's Demon

One of the talented scientists and mathematicians who went through the
doors that Newton opened in *Principia* was Pierre Simon Laplace (1749–
1827). Together with Joseph Lagrange, Leonhard Euler, Carl Gauss, and
others, Laplace developed the mathematical analyses of planetary motions
to a sophistication and precision beyond Newton's. Their work allowed astronomical predictions far into the future, and extrapolations far into the
past.

If all the matter in the universe, and especially that on earth, including
all living things, were composed of particles whose motions obeyed similar
equations, future events could be predicted by a sufficiently comprehensive
treatment. Laplace imagined that

An intelligence knowing, at a given instant of time, all forces acting in nature,
as well as the momentary positions of all things of which the universe con-

sists, would be able to comprehend the motions of the largest bodies of the world and those of the smallest atoms in one single formula, provided it were sufficiently powerful to subject all data to analysis; to it, nothing would be uncertain, both future and past would be present before its eyes.

This came to be known as Laplace's demon; today it would be a super-computer, but whether demon or computer, it represents the philosophy of determinism, which shares with astrology the presumption that the future is already determined by the conditions of the present. Calvinist predestination is a theological version of the idea, which can also be expressed that every-thing that happens, including our own actions, is inevitable, and everything that does *not* happen is impossible. As someone put it: "God could not have operated in any other way." In this view, cause-effect relationships are con-sidered direct, certain, and exact. One might say that physical effects (or changes) are related to causes somewhat as accelerations are related to forces, as implied by Newton's laws.

Today, Laplace's demon is considered an impossibility. As the list of scientific achievements and discoveries grows, so does the list of things shown to be impossible, or achievable only under very special circum-stances: perpetual motion, bodies moving faster than light, heat or time flow-ing backward, living organisms appearing spontaneously, isolated magnetic poles, and so on. Every discovery is a signpost that certain limits and re-lationships hold. But discoveries also indicate what does *not* hold, the de-signs that will *not* work (see Appendix B). Laplace's demon implies that everything can be known, or entered into a computer memory.

Laplace's demon assumes that the information that completely describes and defines the physical universe is not only finite, but also exact. The laws according to which things happen are assumed to be universal and precise; if our scientific theories and laws are imprecise or incomplete, that is pre-sumed to be the fault of the scientists, and of their imprecise measurements.

Yet it is Nature, not man, who abhors perfection, and shows it in many ways. *First* of all, as we increase the precision of any measurement, we encounter noise in each stage of the physical process. There is always some cross talk and interdependence between what we are interested in and some of the factors that we are trying to exclude. One source of noise is thermal fluctuations; since temperature is statistical, such fluctuations are inevitable. Another source of noise is the quantization of electric charge (never smaller than one electron); this makes a very small electric current irregular, so that it seems like a beam of shotgun pellets; it is therefore called "shot noise."

We can use a galvanometer in which a very lightweight mirror rotates through a small angle when current is detected. If we reflect a light beam off the mirror, the reflected beam can travel as far as we like before we let it hit the scale on which we make the reading. That is, we can increase the sensitivity and magnification of the galvanometer as much as we please by increasing the "throw" of the light beam. However, the strategem does not work very well, because the noise in the galvanometer increases along with the signal in which we are interested, and the reflected beam has a bad case of the jitters (a form of Brownian motion).

Second, as though the ubiquity of noise were not enough, we are condemned to finite measurement times, which means that the bandwidths and resolutions of our instruments are inevitably finite. Thus, even in the absence of noise, we cannot reduce experimental error to zero. For example, no wave train can be perfectly monochromatic, because it cannot be infinitely long. And no wavelength is infinitely short, so there is always diffraction.

Third, there is always the possibility of human error, the finite possibility that we may be mistaken, at least a little bit. As we become more sophisticated, we decapitalize Truth and write truth.

Fourth, our concepts and mental constructs tend to be more rigid and schematic than the physical reality. When calculations are too precise, they do violence to the innate fuzziness of reality. Physical variables and constants, unlike their mathematical stand-ins, lose their meaning long before a thousand significant figures are reached. Pressure and specific gravity lose their meaning for very small samples, and thus have limited meaning even for large samples. We may mean exactly what we say, but what we say does not have an exact meaning.

Snow White's wicked stepmother asks, "Mirror, mirror, on the wall, who's the fairest one of all?" On what basis can the mirror give an answer? Is pulchritude a precisely measurable variable? How about intelligence? Mass? Age? Temperature? The concepts themselves fall apart, beyond a certain range. A definition always has a finite range of validity.

How high does the earth's atmosphere extend? This is not, strictly speaking, a proper question; there is no correct answer. The atmosphere gets thinner and thinner as we proceed away from the earth's solid surface, and where we choose to say that we have passed out of the atmosphere is arbitrary, depending chiefly on what properties we are interested in. We can, artificially, define two states or modes for our spacecraft: airborne and vacuum. In one we can use wings and turbojets; in the other we must use rocket power. We shall encounter difficulties with this type of logic when we discuss Schrödinger's cat on Quantum Road.

Fifth, we say "The same water never runs under the bridge twice." Meaning, stasis is a mathematical, or metaphysical, idea, not found in the physical universe. If everything is in a state of flux, then what we wish to measure is in flux too. Immutability is a form of perfection, as unattainable as other perfections. No physical process can be perfectly reversible or perfectly repeatable. The ancient philosophers attached great significance to the distinction between Change and Stasis (No Change). Galileo and Newton shifted the significance to the distinction between Uniform Change and Non-uniform Change (acceleration).

As a consequence of all these limitations, our predictions of, for example, eclipses must become less and less accurate as we extend the calculations further and further into the past and into the future. Our data on the motions of the earth and moon are of limited precision, and their orbits are of limited constancy. Like everything else that is physical, they change with time. Laplace's demon should have been exorcised at birth. The only places we find Perfection, Exactness, and Unchangingness are in the concepts that we so define, such as in mathematics. Perfection is always man-made, and simplistic. Laplace's demon was a perfectionist fantasy.

Roads to the Big Questions

Each science tends to recapitulate the same stages, passing through analogous phases. Here is a chronological, or evolutionary, schematic of the stages. Like all schematics, it distorts and oversimplifies, but I think it is nonetheless illuminating.

1. *Anthropomorphic stage.* Natural processes are attributed to beings with purposes like our own. We no longer say that Nature abhors a vacuum, but we speak of a battery as having died, or a volcano coming back to life. In biology, there is still a temptation to assume that every organ and natural process has a purpose that seems useful to us. Magic is an offshoot, and predecessor, of primitive scientific explanations.
2. *Segmentation and Classification stage.* Phenomena and characteristics are systematically categorized, classified, and labeled. In biology, this stage is exemplified by Carolus Linnaeus (1707–1778). Physics was segmented into mechanics, heat, light, sound, electricity, and magnetism, each of which had its special fluids or substances. Gravity was associated with phlogiston; heat was due to caloric; light was due to corpuscles traveling through an ether; electricity was a fluid, or maybe two fluids, one positive, one negative. Newton objected: "To tell us that every species of things is endowed with an occult specific quality by which it acts and produces manifest effects, is to tell us nothing." (The reader may reasonably question if Newton's concept of a force does not fall under this criticism.)
3. *Mathematicalization stage.* Observations and experimental results are unified by theories, equations, laws. Isaac Newton is the symbol of this stage, and Laplace's demon can be seen as a metaphysical caricature.
4. *Ankylosis stage.* Apparently disconnected phenomena are discovered to be interdependent and/or to follow similar laws. Unified theories are

sought. Alongside the growing list of things discovered feasible is a shadow list of things discovered to be incompatible or impossible, for example, the exclusion and uncertainty principles in quantum mechanics. As we learn to simulate and duplicate, for example, cloning, nuclear fusion, synthetic materials, we also learn things that *won't* work.

5. *Where Do We Fit? stage.* Not only biology, but all of science took a great step forward with the theory of evolution of species, first announced by Charles Darwin and Alfred Wallace in 1858–1859. The ideal of evolution has become almost a prototype, or philosophical principle, for scientific theories in every field, and a major factor in the modern zeitgeist. Whether a theory deals with galaxies, societies, languages, molecules, or subatomic particles, scientists always hope that different types and species can be shown to have evolved from each other, or to be different stages of a common evolution. Thus evolution is not only a discovery of science, but also a goal. We could write: evolution = continuity + causation.

Copernicus displaced man's earth from the center of the universe, but Darwin put man himself on a mere branch of an evolutionary tree. And the modern temper did not stop there, but continued searching for answers to the Big Questions ever closer to home. Marx and Freud, two bearded Jewish patriarchs, are like the two faces of Janus, one facing outward, to our society, history, economics, and politics, and the other facing inward to our emotions, motives, behavior, and dreams.

Scientists are always looking for what does *not* change in a changing situation, what some philosophers call its "essence." Scientific laws, like the conservation laws of energy, mass, and charge, are examples. So are equations, because they express fixed relationships between magnitudes that vary.

One of the fighting fronts of philosophizing today is in theoretical physics. The nature of reality, the structure of the universe, the role of man-as-observer, the ultimate meaning of space and time, the domains of causality and chance, were formerly disputed by philosophers and theologians. The ticket for entry into today's disputes on such questions is a degree in physics or applied mathematics.

These questions have not arisen suddenly, or as a result of any single discovery or concept. Science, and culture generally, advanced along many roads.

The secret of success and enjoyment in science is not mathematical ability, nor memory for facts, nor ability to observe, valuable as these are. The decisive element is curiosity, tracking down the real questions. One bitter cold morning, I was on my way to school past the lake, which was frozen over. I saw the ducks standing on the ice, and I wondered, "What keeps the ducks' feet from freezing?'' If you or I were standing barefoot on the ice like that, in an hour we'd be hospital cases.

Life is more complicated if you have an active curiosity, and you are more likely to get into trouble, but it will be more fun. A horse race is exciting because nobody knows, despite diverse opinions, which horse is going to win. Science is exciting, but its main purpose is to de-excite, to take the guesswork out of the horse race. Curiosity, according to the adage, killed a cat, and satisfaction brought it back. The adage has it wrong. We cats stay alive by being curious, having a chronic hunger for understanding.

Now let us take a brief look ahead at the roads we will travel away from the Newtonian view of the universe. Already in Newton's time, a wave theory of light was presented. Whether light consisted of waves or corpuscles was much discussed, with the prestige of Newton weighing heavily on the corpuscular side. The phenomena of interference, diffraction, and polarization were decisive in the success of the wave theory, but these phenomena, described along Wave Road (Chapter 2), were not understood in Newton's day. In our own century, however, we have found that light sometimes behaves as though it were corpuscular, and sometimes as though it were waves; this wave-particle dualism is one of the central concepts of quantum theory (Chapter 6).

Newton did not work extensively with electrical or magnetic forces. Faraday and Maxwell were the principal investigators on this road; they conceived of fields, endowing the supposedly empty space between electrically charged bodies, and between magnetic poles, with real physical properties. From the field point of view, described in Chapter 3, Newton's law of gravity implied gravitational fields between masses. Einstein's general theory of relativity (Chapter 7) brought the idea of gravitational fields to fruition. The field concept admirably supplements wave theories, because a field, whether material or vacuum, is what waves wave *in*, and propagate in.

Forces between large numbers of bodies, or of particles, are difficult, or impossible, to calculate because there are so many equations, and each particle interacts with all the other particles. Even the three body problem, for example, the sun, the earth, and the moon, is intractable; Newton worked on it and became convinced that a closed (exact) solution was impossible. He was right; we use computers to get very good approximations today. In the nineteenth century, recourse was had to statistical methods. The mechanical and thermal behavior of gases is described by statistical methods. Modern physics has discovered that statistical probabilistic methods are also essential on the atomic level, even if only a single atom is involved. The quantum theory is principally concerned with the probability of physical events which occur on the atomic level; Chapter 4 discusses the role of probability concepts in physics.

The new regions (theories) reached by the roads from Newton were all fascinating, but the strangest were relativity and quantum. These are the heart of the modern scientific revolution. Special relativity (Chapter 5) is a marriage of Newtonian concepts with the electromagnetic field theory of Maxwell, and the general relativity theory extended the theory to include accelerating and gravitational bodies.

An anecdote of Niels Bohr and Wolfgang Pauli, two major physicists, illustrates the intellectual ferment of the six roads from Newton. Pauli gave a lecture on his latest theories to a group of physicists headed by Bohr. The lecture was followed by adverse discussion and sceptical questions. At the end, Bohr arose and summarized: "All of us know that in order to construct a theory which will contain the new facts, some old and sure ideas must be overthrown. The feeling of the group here is that your new theory is not crazy enough."

The six roads from Newton do not lead to answers to the Big Questions concerning the nature, origin, and purpose of man and of the universe, but these Big Questions are never far beneath the surface. Consider the simple matter of looking at ourselves in the mirror. Where is the image? There is an optical image behind the mirror; the light energy is focussed on our retina; the image is perceived in our brain. All three images—the virtual, the real, and the mental—are of scientific interest, as well as that we are looking at ourself, i.e., the content as well as the location of the image. As different aspects of scientific problems are considered, and we travel the diverse directions of the six roads, we will obtain a deeper understanding. Then we can return, in the final two chapters, to consideration of some Big Questions. But don't expect final or complete answers. The scientific revolution isn't over.

2
Waves and Their
Differences from Particles

Newton's laws are centrally concerned with mass and with the motions of massive bodies. The success of Newton's system led to its caricature in Laplace's demon: everything that happens, or can happen, in the universe is the consequence of particles obeying mechanical laws. One of the most difficult phenomena for this view to explain, however, was light. Unlike sound waves, light does not require any medium, especially in coming through interstellar space from the stars. Newton believed that light consisted of tiny corpuscles shot out from the light source. But this view was disputed by scientists in his own time, especially by Christiaan Huygens, who wrote that light consisted of a wave motion. Huygens' viewpoint fit in well with the concept of physical fields, which was central to the optical and electrical discoveries of the nineteenth century.

Put three billiard balls touching each other in a line on a billiard table. Shoot the cue ball along that line, so that it hits the end ball squarely. The ball that is hit, and the middle ball behind it, do not move, but the ball at the far end of the line flies off, with very nearly the velocity of the cue ball. What has been transmitted through the two stationary balls? A pulse of vibrating molecules. A signal that carries energy and information, but not mass. A pattern propagating in a medium. The science that evolved from these concepts is called wave physics; it affords an alternative view to Newton's of the physical universe, because nearly everything, including moving particles, can be construed as composed of waves of some kind. The elastic waves that travel through the billiard balls are similar to the sound waves we hear; the light waves we see are a different kind of wave. There are also water waves, which propagate on a surface between two media. The reader may be familiar with other kinds of waves, such as in spiral springs and in musical instruments.

Newton wrote a book on optics in which he reported his discovery that white light could be decomposed by a prism into the colors of the spectrum. Being Newton, he investigated whether the colors were inherently in the light, or were due to the prism; he decided they were in the light. Newton also wondered, when light is reflected at a glass or water surface, what determines which part of the light will be reflected and which part refracted and transmitted? Newton puzzled over this splitting of a single beam into two beams, but not finding the cause, he fudged an answer. When a genius fudges, even the fudge is interesting; he wrote that the incident light "suffered alternate fits of transmission and reflection." In the nineteenth century, the wave theory of light, especially the work of Augustin Fresnel and, later, of James Clerk Maxwell, seemed to give a complete answer to Newton's question, but the quantum theory in the twentieth century threw the question open again. We will discuss the physics of waves along this first road, on the fascinating journey to the quantum theory.

Until the advent of the quantum theory, the wave nature of light was considered firmly established by the observation of two types of phenomena:

1. Interference and diffraction, which involve waves being superimposed on each other, so that they either augment or cancel each other. That is, if two trains of waves intersect each other, their amplitudes add at the places where the waves are in phase, and subtract at the places where the waves are out of phase.

2. Polarization, which involves the direction in which the waves vibrate, for example up and down rather than left and right, if the train of waves is propagating in a horizontal direction. Sound waves in air and in water vibrate back and forth *along* the line of propagation; therefore, such waves cannot be polarized.

Interference and Diffraction

Light travels in straight lines, which might suggest, as it did to Newton, that light consists of tiny particles shot out from its source at high speed. If light were a wave motion, on the other hand, one would expect it to curl a little around obstacles, the way water waves do around a breakwater. Newton looked for such curling, which is called diffraction, but did not find it because light waves have very short wavelengths. So Newton advocated his corpuscular model; he was not very dogmatic, putting his conjectures in the form of queries. Nonetheless, his prestige and authority were so great that this corpuscular model went almost unchallenged for a century.

In 1801, Thomas Young reported his famous double slit experiment, demonstrating that light does travel in waves. Figure 2 is a view from above, as from an airplane flying over water waves; it shows wave fronts, but not the ups and downs of the waves. When we shine light through a pinhole, which serves essentially as a point source of light, the light waves spread out from the pinhole, the wave fronts being circular (actually, spherical) arcs centered on the pinhole. The waves encounter two slits, symmetrically placed, and these are the only places where the light can proceed. So, beyond the slits, we have two sets of waves spreading out, one set centered on each slit. These two sets of waves partly overlap because the slits are close together. There is a white screen for the waves to hit and be seen. Where the two sets of waves overlap on the screen, alternating light and dark lines (interference fringes) are seen. These cannot be explained on a corpuscular model; where the two sets of waves are superimposed, their amplitudes (in the case

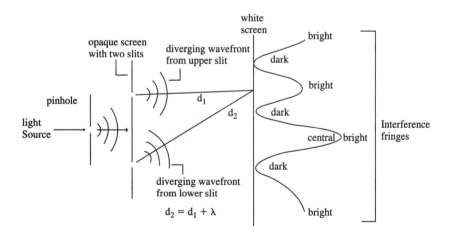

Figure 2   Young's double slit experiment.

of water waves, these correspond to their heights) add or subtract, according to their phase relationship. That is, where the two superimposed waves are in step with each other, they add, and we see a bright fringe of light; where they are out of step, they subtract, which is to say, they add algebraically, canceling like profits and losses, and we see a dark fringe. Figure 3 offers an alternative view of the experiment with a clear view of the fringe pattern created. Dark fringes are strong evidence of light waves, because how else can we explain that we get *less* light at a place on the screen when *both* slits are open than we would get if we blocked one of the slits? Energy is conserved, however, because the bright fringes have double amplitude.

The spacing of the fringes depends on the separation of the two slits, relative to the wavelength of the light. At the center of the screen (optical axis), the distance from each slit is the same, so the two sets of waves will be in phase; we will get a bright fringe due to what is called constructive interference. On either side of the central spot on the screen, the distances from each of the two slits will be different, one being a little closer than the other. If the number of wavelengths from one slit to a point on the screen is one (or two, or three) more or less than the number of wavelengths from the other slit, the two wave trains will be in step, and we will see another bright fringe. But if the distance difference is an *odd number of half-wavelengths*, then the two superimposed wave trains will be out of step, and we will have darkness, or destructive interference. In between these two extreme cases, we will get intermediate amounts of light.

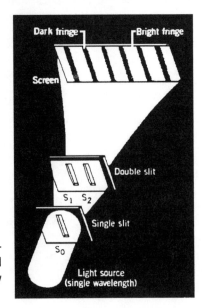

Figure 3 Alternative view of the double slit experiment. (From Cutnell and Johnson, *Physics*, 2nd ed., Wiley, New York, 1991, with permission.)

Young's double slit apparatus is a simple interferometer. We shall encounter nicer interferometers later, using mirrors instead of slits so that we can have much more light. An interferometer is an optical instrument that splits a light beam into two or more beams and then brings the beams together so that they overlap. The trick with all of them, however, is that the superimposed sets of waves must be coherent, that is, have a constant phase relation with each other. It is like platoons of marching soldiers; within each platoon, the men must stay in step. Different platoons may be out of step with each other, but the superposition will be regular as long as they all have the same length of step (wavelength) and the same cadence (frequency). In other words, interference fringes can be seen if distances, or angles, vary between the superimposed beams, but not if the beams are incoherent. For example, we cannot see fringes if we illuminate each of the two slits with light from a different light source, or with light from different parts of the same extended source. That is what the first pinhole is for, to provide a source of coherent light for the two slits. The emitting atoms in a source fire at random times, except in a laser, so we have to arrange the optical system so that it divides and reunites two (or more) sets of light waves emitted from the same atoms at the same instant.

The theory that we have just described was put into mathematical form by Augustin Fresnel. In 1818 Fresnel presented a paper on it to the Paris Physics Society. Siméon Poisson made the following objection: If we place a small perfectly round object, like a penny or a marble, in the center of a light beam, the different parts of the circular wave front passing around the circumference will all travel the same distance to the center of the shadow on the screen, as shown in Figure 4. Figure 5 shows the appearance of the bright spot. So there will be constructive interference from the different parts

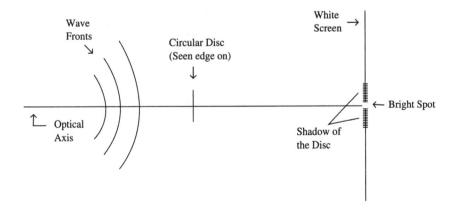

Figure 4   How Poisson's bright spot is created.

of the wave front coming together, which ought to produce a white spot right in the middle of the circular shadow!

Although the phenomenon had not been observed, Fresnel was disturbed by this objection and did not know how to answer it. François Arago, another member of the physics group, came to Fresnel and suggested that they set up the experiment. They did it carefully, and found the white spot. They used a small disk because the wavelength of light is very short, and they wanted a spot in the center large enough to see.

Diffraction plays an important role in both theoretical and experimental physics. The resolution limit of a telescope is usually imposed by the diffraction from the edges of the big lens or mirror. Minimizing the diffraction fringes (which smear a star's image) is one of the two reasons that we make telescopes so big. (The other reason is to gather more light.) The diffraction limitation is also the reason we have to make radio telescope "dishes" even bigger than the mirrors in optical telescopes: the wavelength of radio waves is longer than optical (light) waves, so the diffraction is greater, that is, worse. In microscopes, where we are forced to use very small lenses, we sometimes reduce the diffraction by using ultraviolet light, which has shorter wavelengths than visible light. Still shorter wavelengths can be obtained using electron beams, hence the electron microscope.

The reader is probably familiar with the different wavelength regions of

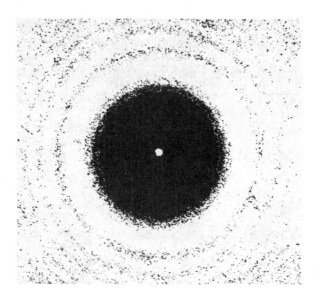

Figure 5 Poisson's bright spot. (From Halliday and Resnick, *Physics*, Part 2, Wiley, New York, 1986, with permission.)

the spectrum, also known as the electromagnetic spectrum. We shall explain why it is called electromagnetic in the next chapter, the Fields Road. Figure 6 shows the major regions of electromagnetic waves. All these waves propagate at 299,792 kilometers per second in vacuum, which velocity is always equal to the product of the wavelength times the frequency.

The critical variable in diffraction phenomena is the angle θ through which the light deviates from a perfectly straight line. This is called the angle of diffraction, and it is measured between the undeviated ray (straight line) and the first bright fringe, counting outward from the center of the diffraction pattern. The equation is

$$\sin \theta = \frac{\lambda}{D},$$

where λ is the wavelength and D is the width of the aperture, such as a slit, or the diameter of the lens or mirror. If $\lambda > D$, diffraction cannot occur; the sine of an angle cannot be greater than unity.

If light and sound are both waves, why can we hear around corners, but not see around corners? The answer has nothing to do with sound waves not being electromagnetic, but is due to diffraction. The sound waves, being on the order of half a meter in wavelength, diffract easily around corners; the wavelengths of visible light, being only a millionth as long, diffract by an unnoticeably small angle. Red light waves, the longest visible waves, are about 0.7 micrometer, and violet light waves the shortest, are about 0.4 micrometer. Blond hair is about 25 micrometers in diameter, brunette is about 30 micrometers (a micrometer is a thousandth of a millimeter or 10^{-6} meters).

For red light passing through a slit 25 micrometers wide, the angle of diffraction is found from the above equation:

$$\sin \theta = \frac{0.7}{25} = 0.028,$$

which gives the angle of diffraction (change of direction of the light rays) as about 1.5 degrees, which is noticeable only under favorable conditions.

Sometimes the apertures are smaller than the wavelengths. What happens then? Waves behave very differently. Radio telescope dishes are made of wire mesh, with open spaces between the wires. This helps to keep them lighter and makes them easier to steer. (They look like gigantic umbrellas blown inside out.) The interesting point is that we don't have to worry about radio waves slipping through the open spaces because the wire mesh size is smaller than the radio wavelengths. People who do not know about limits on diffraction might think that the mesh size, lying *in* the reflecting surface,

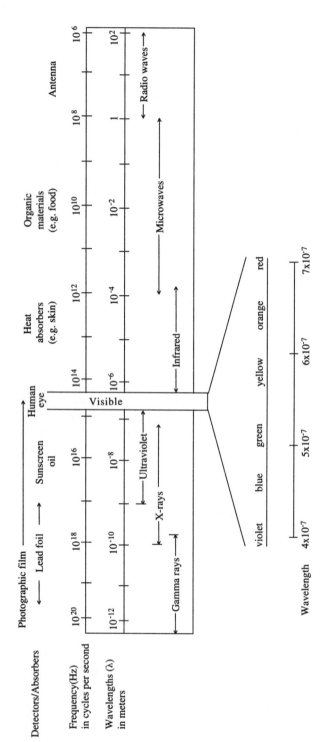

Figure 6 Spectral regions (the electromagnetic spectrum). (Adapted from Cutnell and Johnson, *Physics*, 2nd ed., Wiley, New York, 1990, with permission.)

and *perpendicular* to the wavelength, would have to be smaller than the *amplitude* of the incoming waves, but amplitude has almost nothing to do with it. Amplitude is correlated with intensity, that is, the strength of the wave and the energy it is carrying.

When light waves encounter *opaque particles* that are smaller than the wavelength, the scattering of the waves does not follow the diffraction equation. For example, when sunlight is scattered by tiny particles, such as air molecules and fine smog, the amount of scattering varies inversely as the fourth power of the wavelength. Since blue and violet light have the shortest wavelengths in the visible spectrum, they are scattered much more than the red and orange light. That is why the sky is blue. If the blue and violet light is scattered out of a sunbeam, then the sun will appear to the eye as white minus blue and violet, which is yellow. If you look up at the sun when you are flying in an airplane above 25,000 feet (so that a good portion of the mass of the atmosphere is below you), the sun is much whiter than when viewed from the ground. The sky gets darker as you fly higher; in space, it is black and you can see the stars in the daytime. Red light is scattered only once, while blue light is scattered many times, which is why we see blue sky in all directions and which is why sunsets are red. The one good feature about smog is that it makes the sunsets prettier.

Closely allied with diffraction is interference. Both phenomena exhibit fringes due to the varying phase relationships of superimposed waves, or parts of waves. In diffraction we deal with an edge effect, that is, with waves "curling" around an obstruction, such as the disk that produces a Poisson white spot, or the edge of a lens. What we see on the screen is the sum (or superposition) of different parts of the same wave, arriving by slightly different paths to the same point on the screen; since the different paths have different lengths, the arrivals are not in phase. In interference phenomena, on the other hand, we are not concerned with edge effects, because two (or more) entirely separate trains of waves are superimposed. This is the case with Young's double slit experiment, where each slit transmits its own train of waves. The experiment can be done using a whole row of closely spaced slits; this produces sharper fringes.

There is a nicer way to produce two coherent wave trains, namely, by dividing the wave *amplitude* instead of its *aperture*. We can use partially reflecting mirrors, which reflect and transmit much more light than a slit can pass, and so get much brighter fringes. An oil film on water divides the light that hits it into two beams, one being reflected at the first surface (the air-oil surface), and the other being reflected at the second surface (the oil-water surface), after going through the thin film of oil. (There is a third beam, which goes down into the water and is lost to us.) One of the reflected beams is reflected without going into the oil, and the other goes through it twice, down and back up. If the thickness of the oil is half a wavelength of light, we get constructive interference, because the two reflected beams have a path difference of exactly one wavelength. However, white light has waves of different lengths in it, and the eye sees different wavelengths as different colors. The colors we see in oil films are due to the film causing constructive interference to some wavelengths and destructive interference to other wavelengths. Also, different parts of the oil film usually have different thicknesses, and we view them at different angles, which makes the path differences different. Soap bubbles get their colors in the same way.

The thin film does not have to be of oil or soap; it can be of air. Newton got interference fringes that way, putting a lens that was only slightly spherical on top of a flat piece of glass; this left a thin air space between the two pieces of glass. Figure 7 shows a diagram of the apparatus used to produce the rings, and Figure 8 shows what the fringes can look like. The fringes are circles because the air space is symmetrical about the point where the lens rests on the flat glass. As a matter of fact, we can use any deviation from circularity of the fringes to measure the departure of the lens from perfect sphericity. The fringes, which we still call Newton rings, are like the lines on a contour map, with each ring indicating where the thickness of the air space stays the same. Newton did not find a suitable explanation using his corpuscular model for light.

Low-reflection coatings on glass work similarly to oil films and Newton rings. For low reflection, we want destructive interference, so we coat the glass with a film (magnesium fluoride) one-quarter wavelength thick. Then the beams reflected from the two surfaces of the fluoride film are one-half wavelength out of phase, so we get cancellation. Unfortunately, this works perfectly only for one wavelength, so we choose the middle of the visible spectrum, which is the green, and eliminate the green light. The other colors are eliminated only partially, and the two ends of the visible spectrum, red and violet, are hardly reduced at all, leaving a dim pinkish reflection.

What happens to the energy that is destroyed? This is an interesting question, but it is usually glossed over. The above explanation is given, then the question of energy disappearance is raised, and finally, the inevitable conclusion is that, since energy cannot be destroyed, it must appear as increased transmission of the light. Why isn't the lost energy simply absorbed by the film or the glass? It isn't, but the question is legitimate.

The correct explanation is worth examining because it parallels some of the quantum paradoxes we will presently encounter. There is a transient effect, a stage of the phenomenon that happens very quickly, and then dies out. Consider a wall of water rushing down a dry stream bed during a flash flood. If there is a rock in the way, some water will splash backward at first. But soon the oncoming stream will engulf the back splash, and the water will flow smoothly around the rock. Something like that happens to the light that is reflected from the film during the first picosecond. But the "splash"

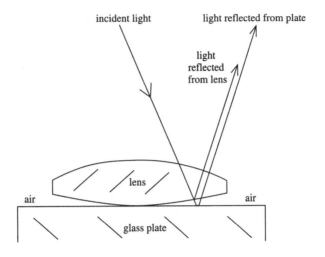

Producing Newton rings – the curvature of the lens surface
is greatly exaggerated

Figure 7 Apparatus for observing Newton's rings. (From Halliday and Resnick, *Physics*, Part 2, Wiley, New York, 1986, with permission.)

dies out, as the oncoming waves sweep all before them. The situation for a precursor wave, which has nothing in front of it, is quite different from the situation of waves coming along later, after steady state has been established. This is why water waves spreading out when a rock is dropped into a pond go only outward; there are no back waves.[1]

What actually happens is quite complicated. "The main body of the signal is preceded by a first forerunner, or precursor, which in all media travels with the velocity c."[2] The precursors are a transient phenomenon; they disappear, and a steady-state situation, including deviation of refracted beams, is established.

An interferometer, like a low-reflection coating on glass, sets up a steady-state interference situation. Figure 9 shows Michelson's interferometer, the most popular type currently used. The entering light beam strikes a semi-reflecting mirror at 45 degrees, which splits the beam into two parts, traveling in perpendicular directions. The mirror is lightly silvered over its whole surface, so that half the light is reflected at 90 degrees, and half the light is transmitted straight through the mirror. Thus we get two light beams that

[1]R. W. Wood, *Physical Optics*, 3rd ed. The Macmillan Co. New York, 1936, p. 8
[2]J. A. Stratton, *Electromagnetic Theory*. McGraw Hill. New York, 1941. Sect. 5.18

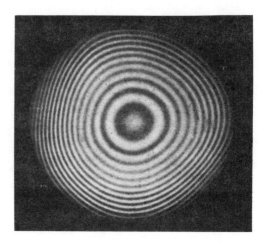

Figure 8 Appearance of Newton's rings (looking down through the lens). (From Halliday and Resnick, *Physics*, Part 2, Wiley, New York, 1986, with permission.)

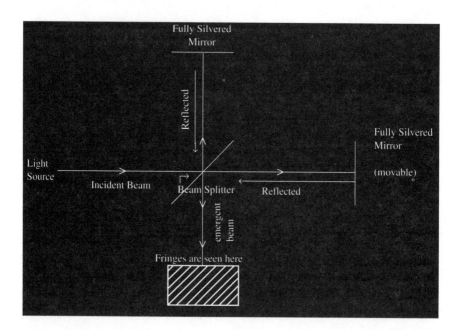

Figure 9 Michelson interferometer. (From Halliday and Resnick, *Physics*, Part 2, Wiley, New York, 1986, with permission.)

are coherent because the amplitude of each wave is split into two twin waves. Each twin beam strikes a *fully* silvered mirror, which reflects all the light back along the same path. The twin beams recombine at the beam splitter. The twin that was reflected the first time is now 50% transmitted, and the twin that was transmitted the first time is now 50% reflected. (This second beam splitting also produces two other beams, which are lost by being sent back toward the light source.) The light that forms the recombined beam in the fourth (downward) direction, as shown in Figure 9, appears to the eye, or on a screen or film, as a pattern of interference fringes. If the two fully silvered mirrors are not quite 90 degrees to each other, what we see is a "contour map" of the air space between the virtual images of the two mirrors. Since the mirrors are accurately flat, we do not get circular or curved fringes as with Newton rings, but straight fringes, since the air space is wedge shaped. The main difference between the Newton ring (and oil film) setup and the Michelson interferometer is that in the latter we completely separate the two beams before recombining them.

One of the fully silvered mirrors can be slowly moved, by a micrometer screw, parallel to itself, that is, closer to or further from the beam splitter. This allows us to vary the path length of one of the beams, without affecting the other beam. So we can introduce whatever path length difference we wish between the two twin beams, before recombining them. Where the path length difference between the twins' trajectories is a whole number of wavelengths, we get constructive interference, and when it is an odd number of half-wavelengths, we get destructive interference. Thus, as we slowly move the end mirror, the fringe pattern moves across the field of view. Keeping count of the number of fringes moving across, Michelson used this method (with some refinements) to measure the wavelength equivalent of the standard meter bar.

There is an interesting limitation to this procedure. As the path length difference increases, so that one wave train (twin) travels many wavelengths farther than the other, the visibility of the fringe pattern washes out, and the pattern disappears. This is because the path length difference approaches the length of the wave train emitted by a single atom. In other words, a twin no longer meets its brother wave train when recombining, but arrives back at the beam splitter after its twin has already returned and left again. As mentioned above, coherence of the superimposed beams is essential, and if a twin recombines with a twin from another pair—a nontwin—interference will not be observable. Playing switch is sometimes possible with laser light, but not otherwise.

Measuring the path length difference at which the fringes fade out gives us a measure of the length of the wave trains emitted by the atoms in the light source. An atom emits light only for a very short time, and by measuring the path length difference at which coherence is first lost (called the coherence length), we get an indication of the length of a wave train twin. Dividing this length by the velocity of light gives us a lower limit for the length of time that it took the atom to give birth, that is, emit the wave train of light that was split into twin beams by the interferometer. This circumstance should be noted, because we will apply it to a major quantum controversy, in Chapter 9.

Holography

One of the nicest applications of interference is holography. Holography not only gives striking effects that are possible in no other way, but also illustrates how much information is lost in ordinary imaging, such as in the eye or with an ordinary camera. In addition, holograms show that information can be coded in sophisticated ways, a sort of super filing system.

A hologram is an interferogram, that is, a photograph of an interference fringe pattern. In a Michelson interferometer, we use two fully silvered mirrors to send the coherent beams back together. Suppose one of those mirrors was an object, say a shiny white chess piece. The light reflected from the chess piece will have different path lengths from different parts of the chess piece, so the interference fringe pattern will be very complicated, and will consist of very fine fringes, so close together that we cannot see them with the naked eye. To avoid having the fringes wash out from excessive path length difference, we use laser light, which greatly increases the coherence length.

Figure 10 shows a set up for making a hologram, and also for viewing it after it is developed. We could use a beam splitter to form the two beams, but it is simpler, as shown in Figure 10, to use two different areas of the laser beam, that is, divide the light by aperture instead of by amplitude. One beam is the reference beam, and the other beam of approximately equal intensity is reflected from whatever we wish to photograph. The two beams are superimposed at a small angle on the fine-grained film; since the two beams are coherent, they form interference fringes. If the two beams are arranged so that they strike the film from opposite sides, we would be making a reflection hologram, such as used in credit cards.

The film, a negative, is developed. It has no image, and looks as though it has been lightly fogged. It is re-placed in the reference beam, although a laser is not essential for viewing. When we look into the beam transmitted through the hologram, as shown in Figure 10, we see the chess piece out in space. There are really two images, but ordinarily we see only the virtual image. The formation (reconstruction) of the image is due to interference of light passing through the photographed fringes, which act like multiple slits in Young's pinhole experiment. Holography is thus a two-part process: first, we photograph an interference pattern, and second, we use the hologram to modify the light and form an image; thus the hologram acts somewhat like a lens.

Ordinary photographs record only the intensity of the light striking the film at each point. A hologram also takes account of the *phase* of the light coming from the chess piece, because the light is superimposed on a coherent reference beam. Thus a hologram contains much more information than an ordinary photograph of the same chess piece. Among other things, a hologram allows us to see the scene in three dimensions. If we move our head while viewing the hologram, the nearer objects in the photograph (per-

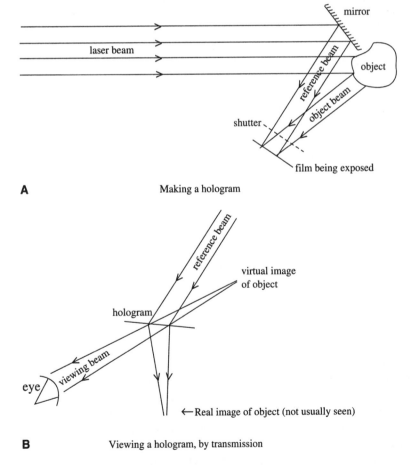

A   Making a hologram

B   Viewing a hologram, by transmission

Figure 10   Making a hologram.

haps there are several chess pieces) will move relative to the more distant objects; this is called parallax. We can actually look around an obstacle and see what, in an ordinary photograph, would remain hidden behind it. The three-dimensionality of the image is not a mere illusion, and does not depend on using both eyes (stereoscopy).

If we cut the hologram in half, or into a dozens pieces, we do not lose any part of the field of view, as would be the case if we cut off a piece of an ordinary photograph. This is because each point on the hologram received light from all parts of the chess piece, as can be seen in Figure 10. In an ordinary optical image, each point in the image (focal plane) receives light that has diverged from a single point in the object; this bundle of rays has been focused by the lenses of the optical system back to a single point, in the image. This is one-to-one mapping, each source point being mapped (imaged) by one point in the image. But a hologram maps the information very differently: each point in the object (chess piece) contributes information (light) to *all* points in the image. A hologram thus records information on an all-to-all basis, that is, holistically. When we cut the hologram into pieces, we still have wave information from all parts of the object on each piece. Of course the devil must have his due: the pieces of the hologram show a less well defined image of the chess piece; it may look as if it had measles (speckles).

The Paradox of the Three Polarizers

Water waves vibrate vertically, but propagate horizontally. Such waves are called transverse waves. Light consists of transverse electric and magnetic waves; the direction of their vibration need not be vertical, but can be in any direction perpendicular to the direction of propagation. The plane containing the vibrations, and the direction of propagation, is a plane of polarization. Usually the electric vibrations are not so neatly contained; light may be unpolarized, or polarized in more than one plane. The present discussion deals only with simple plane polarization.

There are several ways of polarizing a light beam, the easiest being to transmit the light through a crystal of calcite, or through a sheet of synthetic material like Polaroid. Polaroid sheets have microscopic crystals that are parallel to each other, and which plane polarize the light that passes through them, as seen in Figure 11. If a second sheet of Polaroid is also placed in the light beam, and its crystals are parallel to those in the first sheet, most of the light will pass through both sheets. However, if the two sheets are crossed, placed with their crystal axes perpendicular to each other, as shown

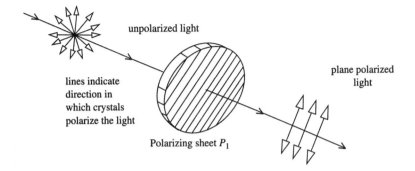

unpolarized light

plane polarized
light

lines indicate
direction in
which crystals
polarize the light

Polarizing sheet P_1

Figure 11   The polarization of light. (From Halliday and Resnick, *Physics*, Part 2, Wiley, New York, 1986, with permission.)

in Figure 12, the second sheet will absorb all the light transmitted through the first sheet; that is, no light passes through the crossed Polaroids.

If the second Polaroid sheet is rotated about the optical axis of the light beam, various fractions of the light will be transmitted, total extinction being obtained only when the two sheet axes are accurately 90 degrees apart. Suppose we cross two Polaroid sheets, so that no light gets through, and then place a third Polaroid *between* them at 45 degrees. Quite a bit of the light gets through! How does it happen that, without touching the two

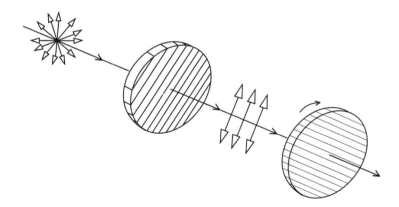

Figure 12   Unpolarized light is not transmitted by crossed polarized sheets. (From Halliday and Resnick, *Physics*, Part 2, Wiley, New York, 1986, with permission.)

crossed Polaroids, interposing a third Polaroid "restores" some of the light? All three Polaroid sheets are identical, and the paradox can be repeated placing them in any order along the path of the light.

Some enthusiasts for quantum theory have written that this paradox of the three polarizers is an example of special quantum statistics applied to light considered as consisting of photons, or corpuscles of light. However, no such special assumptions or statistics are required, merely consideration of what a polarizer can do to a train of transverse waves.

A Polaroid sheet *rotates* light waves, lining them up so that they all vibrate in the same direction. Unpolarized light has vibrations in all directions perpendicular to the direction of propagation. Some light waves need to be rotated through only a small angle to get them in the "right" direction; others will require a greater rotation. The effectiveness of the polarizing action, that is, of the rotation, depends on the amount of rotation required. If only a small angle is required, the polarizing action is very efficient, and nearly all the light is lined up. The efficiency varies as the cosine of the angle. However, if the light vibrations are 90 degrees out of line, then the polarizing medium is unable to swing them around. This is why a second Polaroid crossed with the first Polaroid blocks the light, instead of just re-rotating it. But if another Polaroid is interposed between the crossed Polaroids, then the interposed Polaroid will rotate most of the light that is polarized by the first, and some of this rotated light will be rotated again by the last Polaroid, and transmitted. If we were to use four or five Polaroids, the effect would be similar, provided that the axis of each Polaroid is oriented at an acute angle to the Polaroids on either side of it, the trick being to divide up the work of rotation. Polarization, like interference and diffraction, is explicable as a wave phenomenon, but not readily handled in a corpuscular model.

Total Internal Reflection

A significant and revealing type of wave behavior occurs when light tries to emerge at a steep angle from glass or water into air. The significance of this behavior will become clearer when compared with analogous particle behavior. According to Snell's law, an emergent light beam must be refracted *away* from the normal direction to the emergent surface. But this may be impossible, because the angle cannot exceed 90 degrees, and the sine of an angle, according to Snell's law, cannot exceed unity. It's like the old conundrum: How far can a dog run into the woods? Halfway; after that, he's running out. If the light has to bend more than 90 degrees, it is bending back *toward* the perpendicular. The light tries to come, but if it can't make it, it goes back, that is, it is totally reflected.

This is the best kind of mirror. The light, on striking the reflecting surface, tentatively penetrates through it a short distance, as though it were seeking more glass. If another piece of glass is placed very close by, within a wavelength of the light, some of the light will pass into it.[3] We call this behavior *tunneling* (through the air gap). This overshoot behavior also occurs with particle beams, when the particles encounter a new medium, which, ordinarily, they cannot cross. This is a quantum phenomenon, described in Chapter 6 as the tunnel effect. Tunneling is understandable from the classical viewpoint when it is interpreted as a tentative, or conditional, overshoot. Consider liquid sloshing in a cup or other vessel. If the pulse (wave crest) of the liquid rises above the rim only slightly, the liquid will fall back inside, without any spilling (total internal reflection). However, if the crest encounters blotting paper, or otherwise finds a "favorable home," it will leave the main body of the liquid, that is, "tunnel."

Tunneling is a sort of reconnaissance operation (transient). If the point men find a favorable situation, they signal the whole platoon to come ahead; if the point men find unfavorable terrain, they pull back and rejoin the platoon.

If you have seen binoculars with the cover removed, you may have wondered why the reflecting surfaces of the prisms are not silvered. The reason is that some of the penetrating wave energy will be absorbed by the silver, and the reflection will be less than total. An optical technique that uses this effect, called frustrated total reflection, measures optical properties of materials by pressing them against a glass surface and attempting to get total internal reflection of light.

Fourier Components: Everything Has Waves in It

One of the very useful things about waves is the way they add to, or subtract from, each other, called constructive and destructive interference. There is a wonderful mathematical method that imitates this, developed by Jean Fourier in 1822. Fourier was calculating the temperature in a thick solid slab which is being heated on one side. The heat flows through the slab, so each point in the slab has a temperature which depends on the distance from the hot side and on the time. Fourier expressed the variations of temperature as sums of sine functions and cosine functions. We visualize sines and cosines as waves, but there are no waves of heat or temperature in the slab. Never-

---

[3]A. Sommerfeld, *Optics*. Academic Press. New York, 1954. Sect. 5B and 5C.

theless, the Fourier series (sums of sines and/or cosines) is a powerful way of representing functions, even if there are no physical waves at all.

It turns out that almost any mathematical function can be "decomposed" in this way. The conditions are so broad and generous that math professors sit up nights thinking of functions that can*not* be represented by Fourier series. This means that *we can regard almost any physical process as consisting of waves (periodic processes) added to each other.* In music, we have the fundamental note and the various harmonics; the harmonics are the terms—components—of the Fourier series. Resolving a physical system into its Fourier components provides a powerful analytic tool in nearly all branches of physics. For example, sound waves can be analyzed in terms of the amplitudes of the components of the different frequencies (pitches). A woman's voice usually has larger amplitudes of high-frequency components than a man's voice. The harshness of voices heard over intercom systems is usually due to the high-frequency components having been cut off (filtered out or absorbed) by the circuit, generally to minimize the bandwidth of the channel.

Figure 13 shows two examples of how waves, when added properly, can produce simple shapes (graphs). In Figure 13a, waves are superimposed to produce an approximation to a diagonal straight line. The word "superimposed" means "added algebraically," so that the phase of the waves is taken into account; when a downward amplitude is superimposed on an upward amplitude, the small amplitude is subtracted from the larger, as with profits and losses.

The sum of four terms is shown, each term of the Fourier series being a simple sine wave, with many cycles or periods in it, and each cycle having the same wavelength and amplitude. The figure then shows the result when six terms are superimposed, and then with 10 terms. Each term has a shorter wavelength, and usually also a smaller amplitude, than the preceding waves. The more terms that are added, the closer the resulting curve approaches the desired shape or line.

In Figure 13b, the same method is used to approximate a square wave. Four terms of the Fourier series are superimposed. The method may look clumsy, and the approximations poor, but mathematically it is elegant, and when many terms are added, the approximations become so good that we cannot distinguish them from perfection. More precisely, all the experimental information is preserved. Fourier components really do a good job.

We conclude this summary of wave physics by working out a simple problem. We use light waves, which all have the same velocity c in vacuum (and nearly so in air), regardless of the wavelength or frequency. The wavelength times the frequency equals the velocity, so if we measure the wavelength (as in an interferometer) we can easily calculate the frequency. Conversely, if we know the frequency we can easily calculate the wavelength. For example, the wavelength of green light is half a micrometer. We express the velocity in micrometers per second (to keep apples with apples); it is 3 × 10^{14} micrometers per second. We simply divide this velocity by the wavelength,

$$\frac{3 \times 10^{14}}{0.5} = 6 \times 10^{14},$$

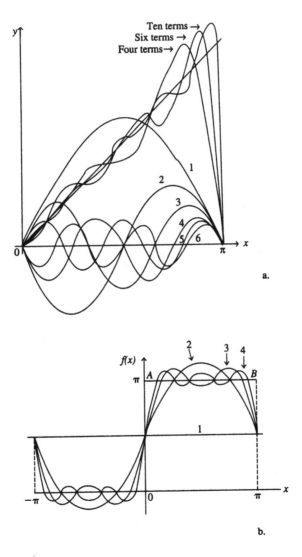

Figure 13   a. Waves producing a diagonal line. **b.** Superposition of sine waves so as to produce square waves.

and get a frequency of 6×10^{14} cycles per second for green light. This means 600 trillion vibrations made by a light-emitting atom in one second. However, an atom usually does its vibrating in much less than a second.

On the next road from Newton, which deals with fields, we will examine the sort of media in which all these vibrations take place. This will not only enhance our understanding of waves, but also prepare us to understand relativity. In addition, it will help us understand many other kinds of forces, especially electric, magnetic, and gravitational forces.

3

Fields: Space Is Not Empty

Newton's laws deal with forces, without dealing with how forces act on distant bodies. When bodies exert force on each other, does that mean that a wave of some kind is propagated from one body to the other? In the wake of Newton's law of gravitation, Cavendish and Coulomb (independently) discovered that the force between two electric charges increases proportionately to the product of the charges and decreases proportionately to the square of the distance between them. The analogy with Newton's law of gravity is good, with charges substituting for masses. However, charges come in two types, positive and negative, and so attract or repel each other, while masses can only attract.

We feel intuitively that two bodies can push each other, like rude people in a crowd, because two bodies cannot occupy the same space at the same time. But how can bodies *pull* each other, attracting each other at a distance? How does gravity work? Newton searched for an explanation, but did not find an answer. An answer did not come, in fact, until 1916, in Einstein's general theory of relativity. But long before the gravity puzzle was cracked, the forces between electric charges, and between magnets, were investigated. The law for two magnets, like that for two masses and two electric charges, is also an inverse square law. This suggests that each of these phenomena— and we can include light because illumination also falls off inversely as the square of the distance—involves a geometric property of space. (This idea was discussed in terms of the surface area of imaginary spheres in Chapter 1.) Physicists use the term *field*, meaning that space has physical, as well as mathematical, properties. A field, then, is a "physicalized" space.

Fields are a central concept of physics; the physical universe can be conceived as consisting of fields of different types and strengths. As we shall see, even probabilities are considered as forming fields; what is more, in quantum theory, there are what amount to probability waves, and they interfere, constructively and destructively, with each other!

The ancient Greeks knew that certain materials, like fur and glass, attracted or repelled other materials when they were rubbed. Amber shows this effect particularly well, and the Greek word for amber gave us our word "electron" and "electricity." But these electrostatic phenomena were treated as parlor curiosities, not scientific effects to be studied. This neglect changed near the end of the eighteenth century, when electric currents, rather than static charges, were discovered. Luigi Galvani, an Italian biology professor, touched two wires immersed in a special liquid, to the leg of a dead frog. The frog jumped, and Galvani probably jumped too, because the frog was very dead. Thus was discovered that muscles respond to electric current conducted by nerves. The liquid used was what we today would call a battery fluid, or an electrolyte.

Benjamin Franklin was the foremost scientist in the American colonies. He invented bifocal spectacles, and discovered, by his famous and dangerous experiment with a kite in a thunderstorm, that lightning was electrical. If an educated Greek from ancient Athens had visited Franklin in Philadelphia, Franklin could have shown him the latest technology in many fields, and the Athenian would have been able to understand almost everything. But if Franklin himself would come back today, only one-tenth the time leap that the Athenian had to make, Franklin would be completely bewildered by what you and I could show him. The biggest technological differences between our society and Franklin's are due to what has been done with electricity.

Franklin wondered whether electricity was due to an electric fluid, or whether there were perhaps two electric fluids, one positive and one negative. He opted for a theory of one fluid, with surpluses and deficiencies to explain the charges; he also decided that electric current flowed from plus to minus. Franklin's first guess was a good one, but his second guess was less fortunate, since electrons, having minus charge, actually go from the minus terminal (the cathode). However, today we can also make currents of positively charged particles—the second "fluid"—and these currents flow in the opposite direction.

In 1800, Alessandro Volta made the first real batteries, which are sometimes called voltaic cells. This gave scientists the ability to generate electric currents on a continuous basis. The next discovery brought magnetism into the picture; magnets were known to the ancients, and the ancient Chinese invented the compass, but magnetism was conceived as entirely separate from electricity. In 1820, Hans Oersted, a Dane, found that when a wire is carrying an electric current, the wire is surrounded by what we call a magnetic field. More particularly, Oersted found that when a compass needle is placed near a current-carrying wire, the needle turns so as to be tangentially (not radially) perpendicular to the wire. Today, we say that the magnetic needle aligns itself with the magnetic field surrounding the wire.

This is a fundamental fact of nature, and is a favorite demonstration in high-school science classes. You put a piece of glass or cellophane over a wire loop, and attach one end of the wire to a terminal of a battery. Then you sprinkle iron filings on top of the glass, over the wire. When the other end of the wire is attached to the other battery terminal, closing the circuit so that current flows through the wire, the iron filings come to life. They stand up on end, arranging themselves like a marine's haircut. When the circuit is opened, so that current no longer flows in the wire, the filings collapse again into a pile of gray sand. The reason the filings stand up is that the moving electrons in the wire magnetize the space, so to speak, around the wire. What came to be called *lines of force* of the magnetic field form circles around the current in the wire, and the iron filings align themselves along these lines of force. On a much bigger scale, that is what a compass needle does in the magnetic field of the earth.

Oersted did not explain his experiment in these terms, but explanations were not long in coming. The news of Oersted's discovery spread rapidly among European scientists. André Ampère, in France, was excited, and in two weeks worked out Ampère's law, expressing the strength of the magnetic force in terms of the strength of the electric current and the distance from it.

Michael Faraday, in England, was the first to think in terms of fields and of lines of force within fields. Faraday wondered if the magnetizing effect discovered by Oersted was reversible or reciprocal; if you started with a magnet, could you produce an electric current in a wire? Faraday experimented, and in his notebook he describes how he placed wires near magnets, looking for current in the wire. No luck. As he moved the apparatus, however, he noticed that there was a brief pulse of current, but the current immediately fell back to zero. The missing ingredient was *motion!* Faraday rigged up a wheel with which he could rotate a loop of wire between the poles of a permanent magnet. The world's first electric generator! More than 99% of the electricity with which we power our society today is generated using Faraday's discovery. Faraday also reversed the procedure, running an electric current in a fixed magnetic field, so as to produce motion of the wire carrying the current. The world's first electric motor! The reader may be interested in counting the number of electric motors in a typical home.

Faraday came from a poor family and had very little schooling, so his mathematics was weak, but he made up for this with remarkable physical intuition. He interpreted his experiments in terms of fields between the wires and the magnets. He imagined these fields as under stress, due to the observed electric and magnetic forces. The directions of these forces could be described as lines of force, with the stresses corresponding to the magnitudes of the forces. In the experiment with the iron filings, you could make the lines of force visible. Faraday noted that the lines of force behaved as though they were under tension, trying to shorten themselves, that is, to contract longitudinally. The flow lines of an incompressible fluid show similar characteristics. As we shall see in this chapter with Fermat's principle of least time, and again in Chapter 9, physical processes have a tendency to take a shortest or easiest path through the appropriate field.

Though Faraday's intuition was correct, it was left to another scientist to work out the mathematical rules, that is, to describe the interactions of electric and magnetic fields by means of field equations. James Clerk Maxwell, a Scot working in England, decided to put Faraday's work into mathematical form. Maxwell was born to well-to-do parents, had a first-class education, and was a brilliant mathematician. He is perhaps the most important theoretical physicist between Newton and Einstein. He published his work on electricity and magnetism in two volumes, in 1873, and they have probably had more effect on physics than any other book, except Newton's *Principia*.

The heart of Maxwell's theory is contained in four equations, which give the relationship between electric and magnetic fields. Ampère's law is included as a special case. The variations of field strength in space are related to the changes in field strength in time. The creation of magnetic fields by changes of electric fields, and vice versa, are expressed in mathematical terms, making it possible to calculate the patterns of the lines of force in the fields. Magnetic lines of force always form closed curves, because a north magnetic pole is always "married" to a south magnetic pole; neither can exist in isolation. Electric lines of force, however, need not be closed; they can diverge from a single charge, such as an electron "unmarried" to a proton. Maxwell's equations can be found in physics textbooks. One more equation was needed, giving the force on a charge moving through electric and magnetic fields. This equation was supplied by Hendrik Lorentz some years later. With these five equations, and some mathematical dexterity, one can, for example, design antennas for television reception, and much else.

An important consequence of these field equations is that they can have wave (periodic) solutions, such as the waves discussed in Chapter 2. That is, the fields that Faraday and Maxwell studied can have certain kinds of waves propagate in them. Further on in this chapter we will discuss fields without waves, but first let us relate fields to the material of the last chapter.

We can picture how waves are radiated by considering a simple transmitter, consisting of a single electrical condenser. This condenser is a pair of parallel metal plates, with electric charges on each plate. The charges produce an electric field between the plates. If one plate is charged positively and the other is charged negatively, their attraction is "felt" across the gap between the plates, even if it is a vacuum. Now suppose we connect the plates into a circuit with an alternating current (AC). After one-half cycle of alternation, the charges on the plates will be reversed. But this means that the lines of force will now run in the opposite direction; we have changed the electric field between the plates. It is not necessary to actually move a charged body (wire or plate) in order to change its associated field; reversing the charges on the plates has an effect similar to that of the current flowing in Oersted's wire in that it generates (creates) a surrounding magnetic field. But *creating* a field means *changing* a field; the new magnetic field generates (creates) a second electric field. This second electric field, as it builds up, generates around itself another new and changing magnetic field. So on and on, alternate electric and magnetic fields keep generating further fields, and they extend farther and farther from the condenser.

Since they involve both electric and magnetic fields, we call the generated signal *electromagnetic*. What is its velocity of propagation, that is, how fast are these alternating fields transmitted through empty space? You guessed it; the velocity of light, nearly 300,000 kilometers per second. Maxwell calculated the velocity, using previously measured electrical and magnetic properties of empty space, and opened up a new era in physics.

The frequency of the alternating current determines the frequency of the electromagnetic waves emitted. But the velocity, in empty space, remains the same, regardless of the frequency. So if we use low frequencies, we get long wavelengths, that is, radio waves. (Remember, velocity of propagation equals frequency times wavelength.) Higher frequencies give shorter waves, as shown in Figure 6. Maxwell showed that light was electromagnetic, and he also opened the way for discovering waves that were not, at that time, even suspected—for example, x-rays and radio waves.

The existence of a field with waves propagating through it allows us to dispense with the concept of action at a distance. Newton's system could not accommodate the question: If an asteroid crashed into the moon, thus adding substantially to the mass of the moon, how much time would elapse before the increased moon mass would be felt on earth, for example, in our tides? Today we think that the change in gravity force probably would travel through the space between the moon and the earth with the speed of light, taking about $1\frac{1}{4}$ seconds. Gravity waves have not been measured, but scientists like to keep things simple, using Occam's Razor. We are especially anxious to avoid assuming that a body can influence a distant body instantaneously, by action at a distance, because that is difficult to explain as a scientific process. Fields give us an opportunity to explain the velocities of actions, signals, and waves in terms of appropriate physical properties.

To illustrate this idea, let us *try* to send a signal instantaneously, using as our field a matrix of light bulbs, such as the illuminated display board which shows the latest stock prices in brokers' offices, or the matrix of lights running around the New York Times Building in New York, where they flash the latest news. How fast can they propagate the news around the building? We are not moving any mass, just information, in the form of lighted and unlighted bulbs, all of which are in place before the experiment begins. We are moving a pattern, although it is not a wave pattern. At first, we might think there is no upper limit to the velocity we can get. But consider, if a bulb lights up at one end of the board, how fast can we propagate this "news" to the other end of the board? Only as fast as electricity will go. If we control the light bulbs from a central switch equidistant from the ends of the board, we can make the ends light up simultaneously, but that is not sending information from one end to the other, but only from the switch to each end. We shall see on the Relativity Road that an upper limit on signal velocities is fundamental to the laws of physics.

Vector Fields

We have mentioned gravity fields, electric and magnetic fields, and fields of light bulbs. The fields in which sound waves travel always have mass, as do the fields for water waves, earthquake waves, and most other waves— electromagnetic waves being the outstanding exception. But fields do not necessarily have waves. The colors we can distinguish visually can vary in hue (spectral color, running from red to violet). Any given hue can vary in brightness, compared to various darknesses of gray, from white to black. Another axis in the color field is saturation, or depth of color, running from water to dilute dye to deep dye. Thus a given color sample can be described in terms of three numbers, one for hue, one for brightness, and one for saturation. We do not need to talk of waves in such a color field, nor distinguish separations in it by millimeters, that is, by a spatial measure. Our variables are color variables.

Are these color variables sensations—subjective—or are they objectively real? In the visible spectrum, as the wavelength of light changes, the eye sees changes in color (see Figure 6). But what about the colors one sees when pressing the eyeballs? Or the opposite color one sees after staring at a bright color or light (negative afterimage)? The Optical Society of America appointed a committee to write a book entitled *The Science of Color*, but the committee was stymied for years over this question. Eventually agreement was reached on psychophysical definitions. "Color," for example, is considered in terms of non-spatial and non-temporal characteristics of visual sensations, such as those arising from radiant energy stimulation of the retina of the eye.[1]

So there are many different types of fields. Suppose we consider the temperature of a room. We can consider the room as comprising a thermal field, with a certain temperature at each point within it. Now consider that we have a highly electrified ball in the room; then each point in the room has a certain electrostatic potential, which is to say, would exert a certain force on a charge we might place at that point. In the thermal case, we have given all the information when we give the temperature at each point, but in the electrostatic case, we need to know in which direction the electric force is directed, as well as how strong it is. The temperature field is called a *scalar* field, and the electric field is called a *vector* field. At each point in a vector field we need to know two things in order to specify it properly. Accordingly, the mathematical quantities we use, called vectors, are two-headed numbers. Usually the two heads are magnitude and direction, but other pairs are possible. One can do wonderful mathematical operations with vectors, starting with addition and subtraction, and two kinds of multiplication.

As an elementary example of vector addition of a type with which the reader is probably familiar, consider Figure 14. An airplane is flying due north, and its velocity, proportional to the length of the arrow in the figure, is 300 kilometers an hour. It then encounters a cross wind from the west of 30 kilometers an hour. The diagonal of the rectangle is the sum of the two vectors. It is (by the Pythagorean theorem) 301.5 kilometers per hour, at an angle of 5 degrees and 44 minutes east of north.

Maxwell's theory deals with vector fields. When we have waves, we want to know how intense they are (their amplitude) and their direction of propagation. In Newton's gravitational theory, the direction of gravitational force of one body on another, is always along the line connecting the two bodies, so that is a vector field too. But electric forces are quite different. If the charges are in motion, that is, if we have an electric current, the force exerted by this current on an outside magnet is *not* along the line connecting them. Instead the magnet is pushed *sideways*. This force is reciprocal; a wire carrying a current, or an electron moving in a cathode ray tube, is pushed sideways by a nearby magnet. The sideways force is proportional to the velocity of the moving charge. In Newton's mechanics, the force on a moving body is always proportional to the acceleration (not the velocity), as stated in Newton's second law. Moreover, the forces in Newtonian mechanics are always in the direction of the push or pull, that is, always along the line

---

[1]*The Science of Color*. Thomas Y. Crowell Co. New York, 1953. "Introduction–The Historical Background and Evolution of the Colorimetry Report" by Lloyd A. Jones, Chairman, Committee on Colorimetry, Optical Society of America.

of connection; this is similar to the cross-wind velocity vector in Figure 14. Thus, the forces and fields in Maxwell's theory are very un-Newtonian. *Stationary* electric forces seem Newtonian, but *moving* electric charges change the whole ball game. It is this new ball game that Maxwell's equations govern, and the field in which these forces operate is the electromagnetic field.

So it seems that "empty" space is not really empty, but has physical properties that affect the forces operating in it. In 1854, the mathematician Riemann presented a paper entitled "On the Hypotheses Which Lie at the Foundation of Geometry." It was radical doctrine at that time because it took some "authority" over empty space away from Euclid, Descartes, and other mathematicians, and shared it with physicists. We shall see on the General Relativity Road what Einstein did with that opportunity.

Waves fit so neatly into the fields described by Maxwell's equations that even if there were no other uses for fields, the field concept would be justified. However, fields have other convenient characteristics. A field can be conceived as analogous to a contour map of a hilly terrain; on the contour map, the lines of equal elevation above sea level (or distance from the center

Figure 14   Vector diagram for airplane in cross wind (not to scale).

of the earth) are shown as contour lines. Where the slope of a hill is very gentle, the contour lines are far apart, since each line represents a different, but constant, elevation. Where the hills are steep, the contour lines crowd together. At a cliff, the contour lines may be on top of each other. Regarding the earth as surrounded by a gravitational field, the contour lines are lines of equal gravitational potential, or equipotential lines. Lines of force, like those discussed above, are perpendicular to the equipotential lines of a field.

The closeness or separation of equipotential lines thus represents the steepness of a slope, or the rate of change of the field potential. This rate is called the gradient of the field. On a ski slope, the beginners look for small gradients, while the experts enjoy the maximum gradients. Water tends to run downhill along lines of maximum gradients, which is a path of steepest descent, or shortest time. Such paths are called geodesics. Many physical processes can be represented in appropriate fields as following geodesic trajectories. In Chapter 9, geodesy is discussed in connection with causality.

One of the best known examples of geodesic behavior is known as Fermat's principle of least time. When light goes through an optical system, it finds the path of least time, taking short paths in glass and water, where light travels slower, and longer paths through air, where light travels faster. We tend to think of light as a wave phenomenon, but Fermat worked out his principle in the seventeenth century, before the wave theory of light was established. Actually, Fermat's principle does not imply that light is either a wave or a corpuscle; it only says that a light signal gets from one place to another by the quickest route, no matter how complicated the obstacles in the optical field.

The problem for the physicist (I can't speak for the light ray) is like that of figuring the quickest way of reaching a person drowning down the beach from you. If you go in a straight line, you will swim further than necessary, and you swim slower than you run. So you want to run down the beach to almost, but not quite, the point where you are nearest the drowning person, and then plunge in and swim.

Fermat's principle of least time is usually derived in the textbooks from the law of optical refraction, Snell's law. However, we get a deeper insight into physical processes if we reverse this procedure, assume Fermat's principle as an application of a general geodesy principle, and derive Snell's law, and the law of reflection, from it. This is done for a simple case in Appendix F.

Fine Structure of Fields

We have mentioned that fields can help us understand waves, and also geodetic paths. But what about the fine structure of the field itself? Are fields perfectly smooth and continuous, like the mathematical functions that Maxwell and others use to represent them, or are they discontinuous, grainy, and quantized? This question appears in various forms along various roads.

There are at least three aspects of field graininess in fields which are physically significant. First, the amount of information that can be stored in a field, and that we can hope to extract from measurements, is limited by the graininess. This is similar to the problem of getting very fine resolution in a photograph, or on the retina of the eye. We call the ability to distinguish fine detail *resolution*, and describe instruments like microscopes as having a certain resolving power. The diffraction limits on the resolving power of microscopes and telescopes are discussed on page 30 as depending on aperture widths and on the wavelength of the light used. The finer the details we can distinguish, the more information we have. Why can't we simply increase the magnification of our instrument as much as we like, for example, use several eyepieces one behind the other, and get greater and greater resolution? Magnification is easy, but resolution is limited by the apertures and wavelengths used.

The distinction between magnification and resolution in optics is analogous to the distinction between amplification and fidelity in an audio system. A person using a hearing aid is familiar with what happens when he turns up the gain (amplification) because the speaker talking to him has prunes in his mouth. Now the voice sounds much louder, but the speaker has *bigger* prunes in his mouth! In astronomy, sometimes when we increase the magnification of the star image, we get only a larger smeared star. A blow-up of a photograph is only as good as the graininess of the negative.

Let us consider a field which is deliberately made grainy. Photographs printed in newspapers and magazines have to be modified so that they can be reproduced in ink on a printing press. It is not practical to print light and dark areas by controlling the amount of ink on different areas of the same printing plate. So we use halftone dots, that is, make the plate (field) discontinuous. We produce various shades of gray, from almost white to black, by varying the spacing between the halftone dots. Each dot prints the same amount of ink on the paper, but if the dots are far apart, the eye sees the area as almost white. If you look at a picture in a newspaper through a loupe, you will see the little dots. However, the picture will not appear clearer, or easier to recognize, because the additional information you are seeing with the loupe is only halftone dots. In this case higher magnification is not only no help, but may actually make things worse, and increase your uncertainty about some of the details. A dramatic way to show this is with the halftone picture of a printed sign, when the letters are very small. You may be able to read the sign *without* the loupe, but not *with* it. The halftone dots are high frequency Fourier components (discussed in Chapter 2) in the image, but the information of interest is contained in the low frequency components. One can get a similar effect by examining the image on the TV screen through a loupe.

A second significant aspect of field discontinuities is that they affect the physical processes which occur in the field. Quantum theory is built around a fundamental grain size, called Planck's constant; this will be discussed along Quantum Road.

A third significant aspect of field discontinuities is that the probability of finding what we are looking for in a field, or of getting a precise value of the field at every point, depends on the graininess, coarse or fine. Usually, we cannot tell anything *within* a grain, and the grain size determines our experimental error, or the probable error. Graininess is related to probability in physics, because it limits our knowing how true a picture or measurement we have. Probability plays an important role in the physical view we take of the universe. The next chapter travels Probability Road; we will encounter other portions of the road in the final chapters.

4

Probability: What Does It Measure?

Einstein maintained that God does not play dice with the universe, meaning that probability calculations cannot give a complete scientific description of physical processes. The role of chance, indeterminism, or chaos in physical processes is at the heart of the intellectual revolution churned up by modern physics, and especially by quantum theory. When we calculate the odds on something, is that merely an approximation method forced on us by our incomplete information? Is probability merely a set of mathematical rules for expressing degrees of uncertainty? Or, as many physicists believe, is calculation of the probability that a certain atomic event will occur the nearest thing to a cause of that event, that is, a complete description? In other words, is indeterminism a fundamental aspect of physical reality?

A point of view is argued in Chapter 9 that synthesizes these two apparently irreconcilable points of view, a viewpoint that is neither deterministic (like Laplace's demon) nor indeterministic. Along the Probability Road heading toward Chapter 9 we will see some significant scenery. We begin with an apparently pure, simple, and noncontroversial example of a prediction based on straightforward probability principles: What are the odds that I will throw a 7 on my next dice throw?

There are six sides to each of the two dice, and we assume equal chances for each side coming up, giving 6 × 6, or 36 equally probable outcome configurations. Six of these configurations will add to 7, so the odds of getting a 7 are 6 out of 36, or 1/6. Gamblers make money by taking advantage of odds calculated by this method, so we must be doing something right. But for someone who wonders what keeps the ducks' feet from freezing, there are some questions.

1. The calculation seems to be completely independent of experimental data, so what can be its scientific significance? To be scientifically valid, mustn't it be based on trials of some sort? What if we should actually make a large number of dice throws, and get a 7 either more or less often than one-sixth of the time?
2. Doesn't the actual result of my single throw depend on physical factors, such as how hard I throw the dice, and on what kind of a surface? What position are the dice in when I pick them up, how rounded are their corners, how well balanced are they, what is the friction, momentum, skill, etc.? None of these factors was used in making the calculation of 1/6.
3. What is the physical meaning of the "equal chances" that we assumed for each configuration? How can we tell with physical or mathematical certainty that a sequence of dice throws does or does not give a random sample of outcomes?
4. When we make scientific measurements, we say that repeating the same measurement increases its accuracy, or reduces the probable error. Is this true, and if so, how does it work? Does anything like this happen as we throw the dice more and more often?
5. Suppose we cannot distinguish the two dice, one from the other; does that make a difference in the odds? For example, if we cannot tell a 1 and 6 from a 6 and 1, should they count as two configurations, or as only one, in calculating the odds for getting a 7?

Let us consider these questions one at a time. The first is a challenge to logical positivism and the so-called operational principle whereby scientists recognize as real only what can actually be measured. Einstein asked, "Do you mean to say that the moon isn't there if I don't look at it?" The answer favored in this book is that reality is a matter of degree, of quantity of information. The reality, or probability of existence, of the moon is thus not as great when we are not relying on visual information—but it is still convincingly great. Being able to measure something is physically significant, but it is neither a necessary nor a sufficient condition for its reality. Does the tree falling in the forest, where no one hears it, make a sound? To a great probability, yes, but the process is not as complete or as well defined as we sometimes demand. As the baseball umpire says, when a strike he has called is disputed: "They ain't nothin' till I call 'em." We shall meet this umpire again, when we discuss complementarity on Quantum Road.

Suppose we insist on experimental verification of our calculation of 1/6. If I keep throwing 7s successively, in the face of the calculated odds, doubt accumulates concerning the dice, the procedure, and my honesty. In other words, the calculation of 1/6 may have been based on incomplete information; there may be hidden variables. Or perhaps our assumption of equal chances for each configuration (which we call an equipartition principle) may be slightly inaccurate. Or we may be making mistakes in reading the dice (e.g., poor light and/or poor eyesight). Or we calculated wrong, in our counting faces or multiplying them, so the odds are not really 1/6. Or, everything is really OK, but we need a much larger statistical sample, throwing the dice many more times. But does that mean that in subsequent throws the odds are tipped slightly toward getting 7s less frequently? This simple case turns out to be far from simple.

The second question, above, shows it to be still more complicated. Physical factors are essential for physical explanations, but in statistical theories, such as an experimental study of dice odds, such factors are largely passed over. Life insurance companies concentrate on the initial state of the insured, living, and his final state, dead, but ignore how he gets from here to there, that is, what an individual policyholder dies from. This habit of disregarding details does not prevent insurance companies from making a profit, but in scientific theories, like the quantum theory, paradoxes arise when transients are ignored. We recall the discussion of low-reflection coatings in Chapter 2.

The third question asks that the equipartition principle be put to a test. There are many such tests, and they turn out somewhat differently in modern physics than in classical physics. We are still, on the present Road, in classical territory, which includes statistical mechanics, the kinetic theory of gases, diffusion, and the second law of thermodynamics; we will touch briefly on these further down Probability Road.

The fourth question relates to the creation or accumulation of information through repetition of measurement. If an instrument and/or an experimental setup has a limited resolving power, how can long-continued repetition of the "same" measurement transcend that limit? Probable errors of measurement are supposed, by mathematical criteria, to get smaller as the size of the sample, the number of measurements, increases. Is there a limit, and if so, how does it arise? From an experimental point of view, we repeat measurements in order to determine the magnitude of the hidden and/or neglected variables. If we keep on repeating, we not only do not increase our accuracy, but the whole apparatus, if not the experimenter himself, may change or drift. The principled question at issue is: Is there, or is there not, a True or Exact Value of the measured quantity in the sense that there is of the square root of 2? If there is, then perhaps we can continue to approach it, but if there is not, the whole question is moot. Here again, the nature of physical reality rears its head through the mists of probability theory.

The fifth and last of the questions raised above concerns distinguishability. In the Broadway musical, *Guys and Dolls*, Big Jule demands that Nathan Detroit use his (Jule's) dice. Nathan looks at them and says, "But those dice ain't got no spots on them." "That's all right," says Big Jule (who is holding a gun), "I remember where they are." So Nathan throws the dice, and Big Jule hollers "Seven!" Nathan sarcastically asks, "Which is the 6 and which is the 1?" To which Big Jule asks, "What difference does it make?"

Let us take Big Jule's question seriously. Chart 1 shows the configurations that are possible with a regular pair of dice, and also with Big Jule's dice. There are only 21 possible outcomes with the latter, and three of them give a 7. So with Big Jule's dice, the odds for throwing a 7 are 3 out of 21, or 1/7, whereas with regular dice, as we calculated at the beginning of this road, the odds are 1/6. The reader is entitled to a warning at this point; this calculation does not really apply to Big Jule's dice, but it does apply, when indistinguishable electrons exchange places inside atoms, to quantum mechanics. In Chapter 6 we shall encounter indistinguishability again.

Statistical Physics

Let us return to the third question, concerning the validity of equipartition principles, which play a critical role in modern physics. *Why* should each face of a die come up the same number of times? *Why* should head and tails be equally likely in a coin toss? Is there a "forceless force" operating, or a hidden variable, a teleological causal factor? Toward the end of the Field Road, we noted a tendency of physical processes, as exemplified in Fermat's principle, to "prefer" geodetic paths. The exorcism of these teleologies is undertaken in Chapter 9.

Here is an example of how equipartition enables us to predict a result in a fanciful case where we cannot hope to trace the individual events. Newton and his successors were accustomed to set up a separate equation for the position and motion of each individual particle in a system, for example, each planet going around the sun. But when the number of particles becomes stupendous, the task exceeds the capability of even the largest computer, and we must resort to probability methods that require equipartition principles.

We fill a liter bottle with seawater, and somehow dye all the water molecules red. We then dump the dyed water into the ocean and stir well, so that it disperses uniformly through the seven seas of the earth. We then dip up another liter of water, wondering what are the chances of finding one of the red molecules in the bottle? The answer is: We will find more than 20,000 red molecules! The sceptical reader who recalls his high-school chemistry will start with Avogadro's number, 6×10^{23}, the molecular weight of water (18), and some reasonable dimensions for the earth's ocean area and average depth. He will quickly discover that there are many more molecules of water in a liter than there are liters in all the oceans.

Instead of stirring the red-dyed water, suppose we just dump it in the ocean and let nature take its course. We will get the same result, except that we would have to wait longer for the dispersion to become uniform. Naturally occurring dispersion is called diffusion, and it has a predictable statistical behavior. We can observe it if we drop a single drop of dye or ink into a water-filled beaker or bathtub and watch the water gradually cloud up. Diffusion is a very general phenomenon, and occurs in liquids, solids, and gases, at different rates of speed.

Diffusion can be visualized in terms of a man who leaves the saloon very drunk, and starts to walk home. The saloon and his home are in a town laid out in simple square blocks. Every time our hero gets to a corner, he confronts a four-way choice: he can turn around, turn left, turn right, or proceed straight ahead. The equipartition principle holds, so we call it a random walk. Given the number of blocks he would walk home if he were sober, we can calculate how long it will take him to get home when drunk, on the average. If he keeps walking long enough, and does not mind arriving back at the saloon a few times, he will eventually get home.

Chart 1   Dice Configurations

Regular Dice Configurations	Number of Configurations	Dice Reading (Sum of Two)	Big Jule's Dice Configurations	Number of Configurations
(1,1)	1	2	1–1	1
(2,1)(1,2)	2	3	2–1	1
(3,1)(1,3)(2,2)	3	4	3–1 2–2	2
(4,1)(1,4)(3,2)(2,3)	4	5	4–1 3–2	2
(5,1)(1,5)(4,2)(2,4)(3,3)	5	6	5–1 4–2 3–3	3
(6,1)(1,6)(5,2)(2,5)(4,3)(3,4)	6	7	6–1 5–2 4–3	3
(6,2)(2,6)(5,3)(3,5)(4,4)	5	8	6–2 5–3 4–4	3
(6,3)(3,6)(5,4)(4,5)	4	9	6–3 5–4	2
(6,4)(4,6)(5,5)	3	10	6–4 5–5	2
(6,5)(5,6)	2	11	6–5	1
(6,6)	1	12	6–6	1
	36 Total			21 Total

The kinetic theory of gases assumes that a gas consists of a huge number of tiny hard molecules bouncing against each other, and against the walls of the container, in random directions. That is, each of the molecules is in random walk. By the equipartition principle, the pressure on each of the container walls will be equal. The average velocity of the molecules is observed in the form of the temperature of the gas; the hotter the gas, the faster the molecules. If the container has a constant volume, then the gas pressure against the walls will increase with the temperature.

This model enabled Maxwell (the same Maxwell we met along Fields Road) to predict something interesting about the conduction of heat across a gap, or layer, of gas which is contained between two large parallel walls. We heat one wall, and measure how fast heat is carried across the gap (thermal conductivity) by measuring the temperature as it rises at the second wall. If we pump some of the gas out, which would be creating a partial vacuum, there will be fewer gas molecules to conduct heat across the gap, but, on the other hand, each molecule will go further before hitting another molecule. So each molecule, on average, will carry heat (by its kinetic energy, which depends on its velocity) further between collisions. The increase in the length of the mean (average) free path between collisions balances out the decrease in the number of heat-transferring collisions. In other words, the same amount of heat per unit time is transferred by fewer molecules, because each molecule goes further. Thus the theory predicts that the thermal conductivity of a gas will be independent of the gas pressure, and so it is found to be. So if we pump out nine-tenths of the air between the walls of a double-walled container, we do not improve its insulating properties.

However, if you *keep on* pumping, until the mean free path between molecular collisions becomes greater than the distance between the walls, the molecules will hit the walls more often than they hit each other. Then, as we continue to pump out air, the fewer and fewer molecules will transfer less and less heat across the gap. Most of the transfer-collisions will be of molecules hitting walls, not other molecules, so the fewer the molecules, the less heat transferred. This is the way a thermos bottle and a Dewar flask work, with fewer than one out of a million air molecules left from the original (atmospheric) air pressure between the walls.

The Second Law of Thermodynamics and Maxwell's Demon

Maxwell, and also Boltzmann, worked out the statistical distribution of the velocities of the gas molecules. At a given temperature, some of the molecules are moving faster than average, and some of them move slower than average. Thus there is a distribution of velocities, with progressively fewer of the molecules moving very much faster or very much slower than average. It is something like a mortality table, with progressively fewer people dying very much younger or very much older than average. The calculation, whether for molecules or for people holding life insurance policies, is based on the mathematics of probability.

However, when we are dealing with probabilistic phenomena like random walks and statistical distributions, there is a built-in unidirectionality. Things get more evenly distributed, like the red molecules in the sea, spontaneously, but it is quite another matter for them to get sorted out again. In discussing molecular velocities, we stipulated that the directions are *random*. An iron rod at uniform temperature has iron atoms moving in it at different velocities, centered around an average velocity, which corresponds to the rod's temperature. Suppose that I move the rod, like an orchestra conductor's baton, and suppose we neglect air friction. Does my movement of the rod change its temperature? No, because I am not giving the molecules any increased *random* motion; I am moving all the molecules by the same velocity in the same direction. By similar reasoning, a single molecule confined in a box does not really have a temperature, because it is visualized as having a definite velocity and direction at every instant. Randomness requires a large sample of objects, so that the average is physically meaningful. The unidirectionality associated with random phenomena is expressed as the *Second Law of Thermodynamics*.

Suppose I put one end of the iron rod in a furnace. Soon I will feel the heat in my hand. This is due to the diffusion up the rod of the faster molecules. One end of the rod is at a higher temperature than the other. When we have a difference of temperature, such as between the two ends of the rod, we can, in principle, operate a heat engine. A steam engine, for example, works by making the hot steam move a piston. The trouble is that, to keep the engine running, we have to eject the steam after it has done its work, that is, cooled a little, and then inject another load of hot steam. That is why steam engines are not very efficient; they throw away a lot of hot steam. The Second Law says they *have* to do that, and lets us calculate how efficient it can possibly be, under ideal conditions.

Let us use the iron rod, with one end hot and the other end cooler, to run a little steam engine. We can use the hot end to boil water into steam, and the cool end to condense the steam into water again. After a while, however, the hot end will have been cooled, and the cool end will have been heated, so that the rod is at a uniform temperature. Now the steam engine will no longer work, even in principle.

Suppose I prepare to repeat this experiment, but instead of attaching the steam engine to the rod, I put the rod in a thermos bottle, so no heat can leave the rod. When I take the rod out of the thermos bottle, it has a uniform temperature, because the fast (hot) molecules at the hot end have diffused throughout the rod. So I can't run the steam engine at all, even though I haven't lost any heat. Energy has been conserved, but it is no longer available to do work. The natural tendency of all physical systems (not just iron rods) to change so that their energy becomes less available to do work is called the Second Law of Thermodynamics.

The Second Law is one of the ruling principles of science. It is fundamentally non-Newtonian because it is based on irreversible phenomena. In Newtonian mechanics, a system of planets or billiard balls can operate backward as well as forward. In systems dealing with random processes, such as the molecular motion in the iron rod, velocities and temperatures can diffuse, that is, even out, but they are extremely unlikely to spontaneously separate into a hot end and a cold end. In other words, disorder, or degree of randomness, can increase much more easily (with higher probability) than it can decrease.

The disorder of a physical system is called its entropy, and entropy is related to the number of different possible configurations in which the system can exist. At the beginning of this road we considered a system of two dice, and said that it had 36 possible configurations. In the iron rod, there are "skillions" of possible configurations. One way to state the Second Law is: in any closed physical system, the entropy always approaches a maximum. A physical system tends to change toward, and along a path of, configurations of maximum probability. In the case of two dice, we can shoot a 7 by any of six different configurations, so that is the most probable state for the dice.

Carnot showed that the maximum possible efficiency of any cyclical heat engine, such as an ideal steam engine, is given by the ratio of the temperature change divided by the hotter temperature:

$$\text{Max. eff.} = \frac{T_{hot} - T_{cold}}{T_{hot}}.$$

(This formula must use absolute, for example, Kelvin, degrees.) Faraday's way of generating electricity is 99% efficient in practice, but heat engines, such as the turbines used to turn the electric generators, are usually less than 50% efficient. Nuclear power stations are less efficient than that.

There are no known exceptions to the Second Law, and it is a law we would dearly love to break. Wouldn't it be wonderful if we could efficiently take a little of the heat out of the tropical oceans and use it to run machinery? There would be enough energy from cooling the tropical oceans only 10 degrees to meet civilization's needs for thousands of years!

Maxwell tried to find ways to violate the Second Law. Thus was born the famous Maxwell demon. This demon sits in a box filled with gas at uniform temperature. The box is divided in half by a partition, with a little, frictionless door in it. The demon does not do any work (expend any energy) in sliding the door open or closed. The demon can see the molecules; when he sees a molecule moving a little *faster* than average toward the door from the right, he opens the door. And when he sees a molecule moving a little *slower* than average, moving toward the door from the left, he opens the door. All other times, he keeps the door closed. Gradually, the demon sorts the fast molecules into the left half of the box, and the slow molecules into the right half. This means that the entropy of the box has been decreased without work (the expenditure of energy)—a flat violation of the Second Law. The box that started at uniform temperature, now has a hot side and a cold side. Maxwell's demon has caused quite a flap among physicists.

So what if an imaginary, mythical creature *did* violate the Second Law? Scientists are extremely rigid and puritanical about their laws anyway. A scientific law is like virginity; one violation, even by a mythical demon, and it is gone forever. Over the years physicists tried, not too convincingly, to salvage the reputation of the Second Law, but it was Leo Szilard in 1929 who exorcised the demon. Szilard recognized the significance of the demon's dependence on *information* about the molecules. The key to the exorcism was the realization that *information is negative entropy*. In order to see the molecules, and illuminate them against the background, the demon must introduce entropy into the gas, such as by using some sort of flashlight. Brillouin stated that the entropy thus introduced would offset the demon's reduction of entropy, and save the Second Law.

The analysis of Szilard and Brillouin is not accepted by all physicists. A recent anthology,[1] appropriately entitled *Maxwell's Demon*, contains papers that claim that the demon need not introduce entropy to get the information, but that in getting ready for the next cycle, he must do so. The question seems to depend on how we distinguish between introducing information into a physical system, for example, arranging molecules, and extracting information, e.g., shining a flashlight beam.

Either way, the meaning of the Second Law is that entropy increase is inevitable in physical systems, and that as a consequence, all systems are, to one extent or another, irreversible. We cannot exactly duplicate conditions when we retoss the coin or rethrow the dice. If we have a movie film of a person diving into a swimming pool, and we run the film backward, we see the water splash backward, then the diver's feet come out of the water, then he flies upward, feet first, onto the diving board. We know this is impossible, but how do we know this? In Newton's equations, there is no such direc-tionality; the equations work equally well with time running in either direc-

---

[1]*Maxwell's Demon–Entropy, Information, Computing*, edited by Harvey S. Leff and Andrew F. Rex. Princeton University Press. 1990. This is a comprehensive anthology of articles dealing with the Demon.

tion; we can substitute $-t$ for $+t$, past for future. A movie of a billiard shot, or of planetary motion, or of any other system where friction and entropy change are not noticeable, would not show which is forward and which is backward.

When we straighten up a mess, or crystals form out of solutions, or Maxwell's demon sorts molecules, we have apparent exceptions to the Second Law, but they are only apparent. Living organisms grow in orderly cell patterns, which are low-entropy systems, but in so doing they must increase the entropy of the food they eat. Life and growth are only possible by transferring entropy, decreasing it in one region at the expense of increasing it elsewhere. You can't beat the Second Law of Thermodynamics!

Postulates and Meanings of Probability

Classical physics tries to do *too* much, leaving no slack in the Great Clockwork, in the working of scientific law. Physical variables and constants are considered meaningful to any number of significant figures, even billions of decimal places. Imprecision is attributed to our instruments and to our scientists. Physical quantities are defined too precisely, with noise and limited resolving power of our instruments treated as mere nuisances. (Quantum theory goes to the opposite extreme.) When someone demands to know the exact magnitude of something, he either is talking of something the value of which is set by definition (as in mathematics, or on the sales floor) or else he does not understand the physical problem. In short, exact truths are man-made.

Mathematics is man-made, and probability is a branch of mathematics, however often physicists, gamblers, and businessmen may borrow, or adapt, it. Particular difficulty arises from the mathematical concept of infinity, which has no counterpart in physical reality. Very big numbers, though indefinitely big, are not the same animal as infinite numbers. This is the key to Zeno's paradoxes, which are discussed briefly in Appendix D. Doing something as often as you like is not at all the same thing as doing it forever. The validity of limiting cases is not the same in mathematics as in physics.

The first question raised at the beginning of Probability Road led to the following question: if we toss a penny a thousand times, and get 600 instead of 500 heads, does that mean that the odds for tossing heads are a little less than 50% in the next few thousand tosses, so that the results will even out? If not in the next few thousand tosses, then when, if the equipartition principle is valid?

The answer is of use to mathematicians, but not to gamblers: the equipartition principle is strictly valid only for infinite samples. The mathematical odds for tossing a head never change, no matter how badly you are losing, or winning, at any given time. The certainty that you will get an equal number of heads and tails can be proved only in heaven, where tossing can continue without end, and conditions remain unchanged. In the physical universe, the number of heads almost never comes out exactly equal to the number of tails. If the two numbers *do* come out equally for every large sample tried, there are hidden variables. By the same token, if I get 600 heads out of nearly every 1,000 tosses, there are hidden variables.

We see from this example what randomness really means. It refers to a sequential accumulation of data, such as coin tosses or repeated measurements, in which there is no change in the probability of prediction for the next batch of data. If I cannot predict from all my previous coin tosses any better than 50% probability for the next toss, I have accumulated no information from all the past tosses. Similarly, if I calculate the value of pi, or of an irrational number, to 1,000 decimal places, I still am no wiser as to what the 1,001st digit might be. That is why we call such digit sequences random. And that is why no sample is large enough to give us certainty that we are dealing with a truly, or exactly, random process. In practice, in science and engineering as contrasted with mathematics, random processes are those depending on many nearly independent factors.

Probability results should be treated with caution in science. Probability calculations assume, implicitly, that successive trials are completely independent of each other. Every throw of the dice, every deal of the cards (if W. C. Fields isn't playing) is a completely fresh start. Repetition, replication, and duplication of conditions, events, processes, and atomic states are sufficiently similar physically that we can treat them as mathematically identical. Thus, for example, it makes no difference in the dice odds if we tip over a barrel of dice, that is, make all trials simultaneously, or throw the dice sequentially, using a single pair of dice again and again. In physics, we should be cautious about treating time so casually.

Statistical methods are among the least desirable scientific methods because they bypass causal factors. Ehrenfest observed: "If you need statistics to prove that you are right, you are probably wrong." Mark Twain put it: "There are lies, damned lies, and statistics." We can joke about the man who stood with one foot in boiling water and the other in freezing water, and on the average he was comfortable. Or about the school survey that indicated that, on the average, each girl in the school was 0.2% pregnant.

Criticizing the limitations of probability calculations raises the question of why they nonetheless often give meaningful results. The reliability (probability) of a calculated result may itself be subject to probability calculation; the chi squared test is the second term of an infinite regression of uncertainties, each being the probability that the preceding result is reliable. A probability of zero for a certain event need not mean uncertainty but that we are quite certain that it will not occur.

The odds for throwing a 7 in a dice game were calculated at the beginning of this chapter by assuming equipartition between possible configurations, which we counted. The calculation (prediction) of a planet's position, using Newton's laws, does not use equipartition at all. Between the Great Clockwork and pure chaos lies a continuum of calculation methods. Suppose we ask the odds on a unique or unfinished event, such as that Team A, ahead by three runs in the eighth inning, will win the game. We could look up previous cases and count how they turned out. We can also look at the two teams and their individual players, and how they seem to be doing. In other words, statistical methods can form *part* of a calculation, with specific information on the particular situation reducing our dependence on statistical rules. The latter rely on the postulate of equipartition, while the former rely on mathematical methods which presumably parallel causal processes, for example, deterministic equations.

Chaotic systems include systems whose equations do not have exact solutions, or whose initial conditions are too mathematically precise to be physically realistic. Recent research shows that even such systems may exhibit regularities after a time, tending to favor certain ranges of its variables. Quantum phenomena obey special rules of probability. Mathematics is more open to experimental adaptation than is usually realized.

The reader may have seen the following visible demonstration of the validity of probability methods. Balls are dropped, one at a time, from the top of a space between two large, parallel, glass plates. Rows of pegs are arranged horizontally in the space, so that the balls strike pegs as they fall, bounce back upward, hit other pegs, and eventually reach the bottom of the space. It is like a big pinball machine, except that the balls remain where they land on the floor of the space. Gradually, balls pile up on the floor, with most of the balls near the center, vertically beneath the hole from which they entered at the top. The profile of the pile nicely follows the standard deviation curve of probability. What guides the balls so that they conform to this mathematically sophisticated curve, first derived by Gauss? The answer is not teleological, but results from many nearly independent factors determining a single final configuration. The complexity of the causes contrasts with the simplicity of the result: many factors contract into one or two factors. We shall return to this discussion when we consider the role of causality and of probability in physical processes.

So, after all this, what *does* probability measure? It gives a quantitative indication, following arbitrary mathematical rules, of the relative weights of various causal factors, some or all of which are incompletely known. The more complete our information about how a result comes about, the more certain we can be that our prediction will be fulfilled; probability is focused on predictions, and predictions are focused on information. By "information" we mean the same thing as computer people mean; for example, digital information is used in displaying a picture on a TV or computer screen (CRT).

One last point, concerning indistinguishable dice, which God, disregarding Einstein's advice, uses in atomic phenomena. Let us reexamine the question of whether it makes a difference if we can tell which die is which, as for example, a four on the red die and a three on the green die, versus a four on the green die and a three on the red die. We count those as two different configurations, calculating our betting odds. If we are color-blind, the odds do not change, because we do not use the color information anyway. So what *does* make a difference if the dice are indistinguishable?

It makes a difference in the odds if the number of the dice being thrown is not *countable*, that is, if we do not know if the readout (sum) is due to one, two, or ten dice. The idea of indistinguishability conceals two quite different relations: *countability*, where we can, or cannot, count if there are triplets rather than quadruplets on the dance floor, and *resemblance*, where we can, or cannot, tell which of the identical siblings we are dancing with. This distinction plays a role in quantum mechanics, where electrons are called indistinguishable, but are really uncountable because they cannot be definitely located. How can you count if there are triplets or quadruplets if you cannot locate them?

This is more than a semantic distinction. Suppose you draw two balls out of a large sump containing equal numbers of black and white balls. By classical probability rules, there are four equally probable configurations or outcomes: 2 white, 2 black, a black and then a white, a white and then a black. That gives 50–50 odds for a mixed result, and 1 out of 4 for 2 white, and 1 out of 4 for 2 black. But in quantum mechanics, dealing with atomic phenomena, the odds correspond to 1 out of 3 for 2 white, for 2 black, and for mixed, that is, three equally probable outcomes. In rediscussing the double slit experiment, in Chapter 9, an experimental test is proposed, to find which calculation is correct. The issue is already in controversy, as a result of Bell's theorem, discussed in Chapter 8.

5

Special Relativity: Only One Velocity Is Absolute

The theory of relativity arose as a solution to a disparity between two of what were the best established parts of physics: Newton's laws and Maxwell's equations. Newton's laws formed the foundation for a very successful theory of mechanics, described in Chapter 1 as the Great Clockwork. Maxwell's equations formed the foundation for a very successful theory of electric and magnetic fields and how they interact and propagate, described on Fields Road. Newton dealt with forces relating to masses and accelerations; Maxwell dealt with fields containing electric charges and magnetic poles. The disparity between the two concerned whether or not a pulse of light, which is electromagnetic, is like a moving body (a mass) in that its measured velocity depends on the velocity of the observer and/or of the light source.

The Newtonian principle of relativity, especially as implied by Newton's first law of motion, meant that the laws of mechanics must be the same for a fixed observer as for a second observer who has uniform motion in a straight line with respect to the first observer. Neither of these observers can tell by an experiment which of them is "really" moving and which of them is "really" at rest. Notice that there are two restrictions: (1) Only moving bodies (masses) are considered; nothing is said about waves; and (2) The motion must be uniform and in a straight line; no accelerations are considered. Einstein removed each of these restrictions, the first in the special theory of relativity, published in 1905, and the second in the general theory, published in 1916. So, for the special, or restricted, theory of relativity, the question is: Can Newton's first law be extended to include light signals?

Suppose a speedboat is roaring down the lake while we, in a canoe, happen to be paddling in the same direction, parallel, but slower. We will measure the speedboat's velocity as a little slower than would be measured by someone on shore, because the velocity of our canoe is to be *subtracted* from the speedboat velocity. Now suppose that we are paddling our canoe in the opposite direction; clearly, we will have to *add* our velocity to that of the speedboat. So much for speedboats, or bodies in motion.

Suppose the earth is moving toward a star, and we measure the velocity of the starlight as it passes through two shutters, placed in line with our telescope. When the light passes through the first shutter, it starts a timing circuit, and when the light hits the second shutter, it stops the timing circuit, like a stopwatch, so we can read the time it took the light to go between the shutters. Now let us repeat the experiment, but with the earth receding from the star, six months later. Will we measure the star's light as having the same velocity in both cases? Or will we get values different by twice (once added, once subtracted) the velocity of the earth relative to the star? This was the kind of question that precipitated the relativity theory. There were indications, later confirmed, that the measured velocity of the starlight would be the same, with *no* correction for the relative velocity of the earth and the star. If we can't tell whether it is the earth or the star that is moving through the ether (empty space) then motion between them would be purely relative, having physical meaning only with respect to each other.

Unfortunately, we cannot make this measurement on starlight velocity with sufficient precision using equipment currently available. The velocity of the earth around the sun is too small a fraction of the velocity of light. Or, to switch back into our canoe, our canoe is moving too slowly relative to the speedboat to affect our measurement of the speedboat. Maybe some day we will be able to make this measurement with sufficient precision; it would be of particular interest because it would measure the velocity of light on a one-way trip. The methods we use today involve reflecting the light back to the starting point, leaving open the (unlikely) possibility that maybe, maybe some effect on the outward bound beam is canceled by an opposite effect on the return trip.

The very first measurement of the velocity of light was a one-way measurement. It was made by Olaus Romer in 1675, using the regular times at which Jupiter's moons are eclipsed, as seen from earth, as the moons pass around Jupiter. Romer observed that the eclipses occurred slightly ahead or slightly behind schedule, depending on whether the earth was nearer or further from Jupiter. We don't have the solar system's dimensions measured accurately enough to measure the small variation we are now interested in, which is the difference (if any) in the velocity of light for different relative velocities (not distances) between observer and source.

Observers moving relative to each other are referred to as being in separate frames of reference. Our canoe constitutes one frame of reference, the speedboat is a second frame of reference, and the lake and shore constitute a third frame of reference. If we are talking of light waves, we have a big question, which Maxwell was the first to raise: In what frame of reference should we measure the velocity of light? Maxwell's equations do not imply any dependence of the velocity of light on the velocity of either source or observer. Yet, in a Newtonian system, if the star and the earth are not moving relative to each other, the velocity of the starlight should be measured as *less* than if they were approaching each other and *more* than if they were receding from each other. So is the velocity of light analogous to the velocity of the speedboat, or not? The question requires us to consider the velocity of the canoe, and of the speedboat, with respect to the lake, the medium in which the propagating is taking place.

The question is complicated further by the differences between (1) moving bodies, (2) waves *other* than light waves, for example, sound waves, and (3) light waves. Suppose an airplane is strafing us with machine gun bullets; bullets are massive bodies of the sort that Newton's laws deal with. The velocity of the bullets, relative to us, will vary with how fast we run, and in which direction. Also, the bullets will start out faster, relative to us, if the airplane is flying in the same direction as it is shooting in (Newton's law of inertia). A baseball pitcher is not allowed to run at the batter while he pitches the ball. However, if the airplane has a siren, the sound waves will *not* go faster, relative to us, just because the airplane is flying toward us. So we have three very different cases of relative motion: massive bodies, sound waves, and electromagnetic waves.

Here we are concerned only with electromagnetic waves, which is what Maxwell's equations deal with. A number of different experiments were tried to detect the optical, electrical, and magnetic effect of the earth's motion as the earth moves through the ether around the sun. They were trying to understand the puzzle by measuring the velocities of the canoe and of the speedboat relative to the lake. Do light waves propagate through an absolutely stationary ether, the way boats propagate through still water, or the way sound waves propagate through still air, or neither? Can an ether wind be detected?

The experimental results were bewildering. One set of experiments, such as the stellar aberration we will discuss presently, seemed to show the expected amount of ether wind. A second set of experiments, especially the famous Michelson-Morley experiment, showed *no* ether wind at all; these experiments effectively compared the velocities of light beams traveling in different directions; with no effect indicated, the inference was that the ether was being pulled along with the earth, like an atmosphere. A third set of experiments, especially Armand Fizeau's measurement of the velocity of light in moving water, showed a *partial* ether drag in the water; that is, there was a measurable effect on the velocity of light, but not simply the sum (or difference) of the velocities of the light and the water.

Lorentz and Fitzgerald, independently, tried to explain the Michelson-Morley experiment in terms of a new type of distortion of solid bodies and measuring instruments. They conjectured that the motion of the earth was changing some of the dimensions of the interferometer being used to compare the velocities of the light beams; the interferometer arm (see Figure 9) in the direction of the earth's motion through the ether was being shortened (contracted). We still speak of this as Lorentz contraction, but explain it differently. Henri Poincaré suggested that the velocity of light was absolutely invariant, which turned out to be half the solution. The first person to see that "time itself was suspect," as he put it, was Albert Einstein. Einstein meant that intervals of time, and intervals of distance too, are themselves subject to variation, if measured by an observer in a different frame of reference.

Let us summarize the discussion to this point. The problem of reconciling Newtonian relativity of motion with Maxwell's absolute velocity for light involves, surprisingly, just taking those two ideas seriously, that is, elevating them into postulates: The uniform motions of all bodies are relative (as Newton said), *and* the velocity of light is absolute, not dependent on relative motion of source or observer (as Maxwell's equations imply). In order to contain these apparently incompatible statements in a single theory, Einstein had to modify our Newtonian conceptions of time and space.

First, consider the question of the absolute velocity of light. If the velocity of light is increased when a star is moving toward us, and decreased when the star is moving away, then the behavior of double stars would appear very weird. Double stars orbit each other, and so change their direction and velocity with respect to the earth. The light emitted when one of the double stars is moving toward us would get here long before the light emitted when the star's velocity was in the other direction. We might even see the star in several parts of its orbit at the same time. No such weird behavior is observed.

How about if the velocity of light, c, depended on the motion of the *observer*? If that were the case, but motion of the *source* didn't affect the velocity of starlight, then we could distinguish whether the star was moving toward the earth (no effect on c) or whether the earth was moving toward the star (which would increase c). But this is a flat-out violation of Newtonian relativity. So, to conform to what we see of double star orbits, and to retain our postulate of Newtonian relativity for massive bodies, we must adopt the second postulate, namely, that c in vacuum is an absolute constant. This independence of wave velocity from motion of source and observer does not apply to sound, but then sound requires a medium to propagate in; we cannot send sound waves through a vacuum; the astronauts walking on the moon had to talk to each other by radio. Sound waves need their own kinds of "ether," but light needs only the electric and magnetic properties of vacuum.

Physicists were grateful to Einstein for getting them off the hook of the physical properties of the ether. The ether could not be detected. It had no mass or density or structure. It could support electromagnetic vibrations but did not absorb them the way sound vibrations get absorbed by their media. It evidently could pass into transparent materials, like water and glass, without any friction, so that light in these materials could keep on vibrating in ether. Physicists were glad to leave these questions behind.

But a big question remained: If a wave train of electromagnetic waves, a light signal, cannot be regarded as a frame of reference of its own, even though it travels at a finite, measurable velocity, how are we to preserve Newton's principle of relative motion?

Doppler Shift

This formulation of the problem addressed by relativity theory is difficult to understand. It can, however, be approached in other ways, without reference to Maxwell's equations, or to Newton's laws or the relativity of moving masses, until the very end. We can consider optical phenomena involving the velocity of light, and *then* inquire which results are, and which are not, consistent with the relativity principle. That is, we can use trains of waves, called wave trains, to consider if we can physically identify which is "really" at rest of two frames of reference in uniform motion relative to each other. Our answer, of course, will be dictated by experimental observations. The extension of the principle of relativity to include light waves is justified by suitably chosen experimental tests.

We can start with a wave train of sound waves, or water waves, and show how the relativity principle can be preserved only if we consider a third frame of reference, a medium for the waves to wave in. For sound waves we have the air, and for water waves we have the lake. The first two frames of reference are the source of the waves and the observer of the waves. Then we can raise the question that Einstein addressed: what do we do if there is *no third frame* of reference, no detectable medium, as is the case for electromagnetic waves? The absence of a third frame of reference answers a common question about relativity: how does relativity get so closely connected to light waves? What is so special about them? We cannot measure their velocities as they propagate through empty space.

Let us turn our canoe so that we paddle *into* the waves spreading out behind the speedboat, instead of moving parallel to the speedboat. We encounter the waves closer together, so their wavelength is apparently shortened. By the same token, we encounter the waves more frequently, so their frequency is increased. This change of wave spacing, both in space (wavelength) and in time (frequency) due to relative motion between source and observer, is called the Doppler effect, or Doppler shift.

Perhaps the best known example of the Doppler shift is a sound movie of a locomotive bearing down on the camera (observer) with its whistle blowing a high-pitched scream. As the train moves away from the observer, after passing him, the whistle is still screaming, but heard at a lower pitch. This change of pitch, or frequency of the sound waves, is the acoustical Doppler shift. Notice that we have not said anything about a change of *velocity* of the sound waves, only a change of frequency. We could have mentioned a change of wavelength, but since frequency times wavelength equals velocity, that would not necessarily have implied a change of velocity. As a matter of fact, the velocity of the sound waves is not affected by the velocity of the locomotive. The sound waves are generated in air, and it is the mechanical properties of the air which determine their velocity of propagation, for an observer at rest in the air.

Let us calculate the Doppler shift for this case, using simplified numbers. Suppose the sound waves propagate through the air at a velocity (c) 5 meters per second, and the sound has a frequency (f) of 10 cycles per second. If neither the locomotive nor the observer is moving, the stream of sound waves between them will have a wavelength (λ) of half a meter (the propagation velocity divided by the frequency). For the general case, we write

$$\lambda = \frac{c}{f}$$

Now consider that the locomotive is bearing down on the observer with a relative velocity (v) of 2 meters per second. The air in front of the locomotive has waves that are shortened because the source is riding down behind its own previously emitted sound waves. In one second, the locomotive moves 2 meters, so that the ten waves in front of the locomotive are compressed into a distance of 3 meters (5 − 2). This makes the wavelength $3/10$ of a meter. The observer records the frequency as the wave velocity (unchanged at 5 meters per second) divided by $3/10$ meter or $16\frac{2}{3}$ cycles per second (cps). For the stationary case, we had 10 cps, so the Doppler shift is $6\frac{2}{3}$ cps.

For the second case, we make the locomotive stand still on the track, and have the observer move toward the locomotive at 2 meters per second. In one second, 4 more waves (contained in the 2 meters) pass the observer than was the case when observer and locomotive were standing still. Thus the Doppler shift for this case is 4 cps. The Doppler shift for moving source is always greater than the Doppler shift for moving observer. We have for the moving locomotive:

$$\Delta f = f\left(\frac{v}{c - v}\right),$$

and for the moving observer:

$$\Delta f = f\left(\frac{v}{c}\right)$$

These two expressions are *not* the same.

This result is not as strange as it seems (sounds?) because we are really dealing with velocities relative to the *air*, that is, of the medium in which the sound waves propagate. The air provides a third frame of reference. That is, the air can have a velocity of its own, relative to the source and relative to the observer.

The ambitious reader may check the result using the Doppler shift for wavelength, instead of frequency. The moving locomotive gives a decrease of wavelength of $\frac{1}{5}$ meter, the moving observer gives a decrease of $\frac{1}{7}$ meter. Hint: $(f + \Delta f)(\lambda - \Delta\lambda) = c$.

Although there is no change of velocity of the sound waves observed whether the locomotive moves or not, there *is* a change of wave velocity if the observer moves. This difference does not get us into conflict with the principle of relativity because the sound waves have their own, private, frame of reference, the air. The observer sees a change of wave velocity according to whether he is, or is not, moving with respect to the air. In our canoe, we would observe an increase in the water wave velocity, when our canoe moves with respect to the lake.

The situation changes drastically if we substitute light waves for sound waves or water waves. Consider that in the whole universe there is only one star and the earth, a light source and an observer, no third frame of reference. What physical measurement would allow us to distinguish between (1) the star moving toward the earth, and (2) the earth moving toward the star? If there were a rope stretched between them, we could haul it in, but we could not tell which, the earth or the star, was doing the moving. *There is only one motion*, that of approach between source and observer. If we can't use the ether as a separate medium (frame of reference), the two bodies must perforce use each other for determining their relative motion. That is one way of stating the principle of relativity, and if it is to apply to optical cases, the Doppler shift for moving source and the Doppler shift for moving observer must be *identical.* The star sends light to the earth, but there can be no way to assign the Doppler shift in the star's light to the star's motion relative to space.

Astronomers measure the Doppler shifts from different stars; these shifts vary in magnitude, but the interpretation is of relative motion between the different stars, or between each star and the earth. Our calculations of the two Doppler shifts, above, are correct for sound (in still air) but *neither* is correct for light. In order to extend the relativity principle to include light signals, we must make our concepts of space and time dependent on the velocity v between source and observer, as compared to the velocity of light c, that is, v/c.

Scientists strongly dislike dividing their universe into separate domains, with different laws and principles valid in each. Physicists were loath to acknowledge that the relativity principle for mechanics, and for sound waves and water waves, could not apply to optical phenomena, which would have to be assigned to a separate domain, with its own principles of relative motion. Besides, the experimental evidence failed to give a coherent picture of what principles *would* apply for a separate universe of light waves. We prefer an inherent unity in the universe; scientists are monotheists at heart.

Einstein saw that the way out lay in relating measurements of distance—space—and measurements of time intervals—time—to the velocity of light. Newton had written of an absolute space in which all motions could be unambiguously compared. Newton also assumed an "absolute, true and mathematical time, of itself, and from its own nature flow(ing) equably without relation to anything external. . . ." The theory of relativity discards these two absolutes of Newtonian physics, and replaces them with a single absolute, the velocity of light, c. By Occam's razor, that is an advantage. Even more important, all relative motion, if it is uniform, is governed by the same rules; the unity of the universe, as we see it anyway, is preserved.

Stellar Aberration

We can make yet another approach to reconciling the relativity principle with signals (electromagnetic wave trains) that have no frame of reference of their own, and propagate with a constant, finite velocity. We can again begin with mechanical analogies, and again the analogies will face us with a clear-cut contradiction with experiment, calling for abandonment of Newtonian views of space and time. With the Doppler shift, we used wavelength and frequency; this time we use direction and angle.

A person is standing in the rain, which is coming down vertically. He has an umbrella, held straight up. He decides to make a run for it. He tilts the umbrella forward, so that the raindrops in front of him don't hit him as he runs. The angle of tilt is called the angle of aberration; this angle is calculated from the ratio of the horizontal velocity of the runner to the vertical velocity of the raindrops. This ratio gives the tangent of the angle of aberration. If he runs at the same velocity as the raindrops fall, the ratio is unity, and the angle is 45 degrees.

Before we consider the optical analog, using light from a star instead of raindrops, and the motion of the earth instead of a runner, let us look at another mechanical analogy. This time we are on a ship sailing past a hostile shore battery. The shore people fire a cannonball right through the hull of our moving ship, making a hole in both sides. The exiting hole, on the far side of the ship, will be a little nearer the stern than the entering hole, on the near side of the ship, because the ship moves while the cannonball is within the ship. If we were to sight through the two holes, from the far side of the ship, we would see the cannon displaced from the center of our view. The angle by which it is displaced is again the aberration angle, and again its magnitude is calculated from the ratio of the two velocities, the velocity of the ship and the velocity of the cannonball.

Now how about starlight traveling down a telescope tube? If the telescope moves, being on a moving earth, won't the image of the star be shifted? Yes, the angle of stellar aberration, resulting from the earth's velocity perpendicular to the direction of the star relative to the velocity of light, is observed as an angle of about 20 seconds, when the telescope points near the zenith. This seems straightforward, but it is not, because we are comparing a beam of light to a cannonball, or to a stream of raindrops. A light beam does not constitute a frame of reference, but cannonballs and raindrops do. We can track a particular cannonball or raindrop through space and time, but tracking a particular light wave is a very different proposition; in fact, it can't be done.

The definitive experiment was performed by George Airy, the Astronomer Royal at Greenwich, in 1872. He filled a telescope tube with water, in which light travels about one-third slower than in air. The aberration angle stayed the same. Airy's result puzzled physicists, but they did not recognize its significance. One can almost hear the future Einstein ask Airy: "How does putting water in your telescope affect the relative motion between the earth and the star?"

The question calls for stepping back from the paradoxical experimental results and taking a deep look at the whole situation. No one, at that time, had checked the equality of Doppler shift for moving observer with the Doppler shift for moving source, but their equality, when light was used, would have presented the same paradox as that obtained from Airy's experiment, or from the "zero ether wind" result of the Michelson-Morley experiment. Einstein saw that the time periods of events measured by observers moving relative to each other should differ, due to the finite velocity of light. In other words, time is relative to the observer's velocity, with the velocity of light providing the standard of how much variation to expect in the time measurements. If an observer moves at only a very small fraction of the velocity of light, which is our everyday experience, time will appear Newtonian. But at velocities exceeding, say, 100,000 kilometers per second, watch out.

Einstein reached this amazing conclusion by considering how we establish the simultaneity of different events, that is, how an observer decides which of two events happens first. The key to all the paradoxes lies in how we measure time periods, how we *must* measure time periods. Suppose that two bolts of lightning several miles apart are seen to occur simultaneously by a man located halfway between them. But another man, also equidistant from the two bolts of lightning, will not agree that they are simultaneous, if he is riding on a train moving toward one bolt and away from the other. This is because he moves while the light is traveling toward him from one bolt, and while the light is overtaking him from the other. This is of great significance, because we measure *all* events between what amount to marks of simultaneity. For example, we clock a runner's time by two simultaneities: first when we click the stopwatch simultaneously with his start, and again when we click the stopwatch simultaneously with his crossing the finish line.

We do much the same thing in measuring distances; we put the end of the meter stick alongside one end of an object, which is the spatial analog of simultaneity, and note the mark on the meter stick, which lines up with the other end of the object being measured. The Lorentz contraction, in space, is a consequence of the dependence of time measurements on the velocity of the observer with respect to the velocity of light. Since the velocity of light is the same for all observers, regardless of their own velocities, the spatial measurements (distances) that different observers will make will vary with their time measurements. Distance is velocity multiplied by time; if time differs, and velocity does not, then distance must differ.

The big step in building the theory is to calculate the lengths and time periods an observer in one frame of reference (his rest frame) sees in a *second* frame of reference, moving relative to his frame. This is accomplished by transformation equations, usually called simply *transformations*. The same lengths and times are observed by an observer in the second frame, looking at things in the first frame. Thus, the principle of relativity is preserved, because there is no way either observer, or anyone in a third frame, can tell which observer is "correct"; each sees times and distances distorted in the *other's* frame of reference. Time and distance are relative to the observer; each picks the more convenient frame for himself, and considers that he is "at rest." We like to measure most things assuming that the earth is at rest. We think we prove that the earth is not "really" at rest because we can observe its rotation. However, in his General Theory of Relativity, Einstein undermined that "proof."

Lorentz derived the transformations before the relativity theory, and they are still called Lorentz transformations. Einstein derived them in a different way, based on the ideas indicated above. Derivations can be found in many physics books. The characteristic expression of the transformations, almost the earmark of relativity, is

$$\sqrt{1 - \frac{v^2}{c^2}}$$

where v is the relative velocity of the two observers, and c is the velocity of light. The two observers agree on the magnitude of v, but not on which of them is doing it. They agree on c in all cases. This expression always has a value between zero and one; it is worth careful examination. It is highly nonlinear; for the ordinary velocities of everyday life, such as airplanes, bullets, baseballs, and so on, v^2/c^2 is so much smaller than 1 that the relativistic effects are not observable. But as the relative velocity of the reference frames, v, approaches the velocity of light (nearly 300,000 kilometers per second), the effects become enormous. If the velocity of light were infinite, there would be no need for relativity theory; Newton's equations would apply in all cases.

The relativity transformations lead to happy resolution of all the experimental paradoxes that have been mentioned. Questions about the physical properties of the ether disappear, because no ether is assumed. The price that has to be paid is the simplicity and elegance of the Newtonian system.

Summary of the Main Results

1. When a spatial dimension (distance) in one frame of reference is measured from another frame of reference, moving uniformly in a straight line with respect to the first, the spatial dimension is *shortened* by the factor $\sqrt{1 - v^2/c^2}$. This is called the Lorentz contraction. It does not depend on whether the distance is measured in vacuum, air, water, steel, or whatever.
2. When a time interval in one frame of reference is measured in another frame of reference, moving uniformly in a straight line with respect to the first, the time interval is *lengthened* by the factor $\sqrt{1 - v^2/c^2}$. This is called time dilation; it does not depend on what kind of clock you use, for example, count heartbeats, measure beard growth, use an atomic clock,

and so on. Time itself varies with velocity. This is a tough concept for physicists, too.

3. When the inertia of a mass in one frame of reference is measured from another frame of reference, moving uniformly in a straight line with respect to the first, the inertial mass is *increased* by the factor $\sqrt{1 - v^2/c^2}$. This is called the relativistic increase in mass; it is implied by the relativistic changes of time and length, plus the law of conservation of momentum. It figures importantly in particle accelerators, when the particles are moving at a significant fraction of c. It also affects the top speed we can give to spaceships; if we want a spaceship to travel at close to c, it may need a fuel tank the size of the earth.

4. Energy is equivalent to a very diffuse form of matter (mass), and mass is equivalent to a very concentrated form of energy, according to the famous equation $E = mc^2$. This means that the conservation law for energy and the conservation law for mass are not true separately, but only when they are merged, that is, the *sum* of energy and mass is constant in any closed system. The energy of an atomic bomb explosion comes from converting some of its mass directly into energy (see Chart 3). This is an altogether different process than a chemical explosion.

5. No massive body can accelerate from an ordinary velocity to a velocity equal to, or greater than, the velocity of light. The expression becomes imaginary if v exceeds c. (Also, the mass passes through infinity.) These are both physically impossible, which means that "spooky" instantaneous action at a distance, or the transmission of information by any signal faster than light, is impossible. Bell's theorem, discussed in Chapter 8, throws doubt on this conclusion.

6. Magnetism is not a force separate from electrical force, but a consequence of electrically charged particles (usually electrons) in motion. Iron is magnetic because electron motions are aligned in grains of iron atoms. Einstein's first paper on relativity was entitled, "On the Electrodynamics of Moving Bodies"; in the first paragraph he wrote ". . . the reciprocal electrodynamic action of a magnet and a conductor . . . depends only on the relative motion." That is to say, Einstein objected to the conventional distinction between how a current is generated in a conductor (a) when the conductor is moved relative to the magnet, and (b) when the magnet is moved relative to the conductor. This was Einstein's starting point for the whole business. Relativity theory greatly simplifies the explanation of magnetic forces, in terms of Lorentz contractions. (Occam's razor again.)

7. The interrelationship between spatial distances and time durations leads to formulating mechanics in terms of four dimensions (space-time). Space and time are not interchangeable, but they are treated equivalently in relativistic equations. A physical theory consistent with relativity must preserve this symmetry. Maxwell's equations do this, but the Schrödinger

equation, central to quantum theory, does not.

Much nonsense has been written about being "catapulted into the fourth dimension" and about time warps. The idea of four dimensions can be seen from considerations of traffic control. Different cars can use the same values of x and y coordinates (positions) at an intersection, provided they do not also have the same value of their time coordinate. If all three values are the same, they have a collision; traffic control thus involves two spatial dimensions, plus time. The air traffic controller at an airport must also consider the altitudes of the incoming planes, therefore he is working with four dimensions.

8. When an observer sees two reference frames moving with respect to himself and also to each other, all in the same direction, the two velocities do not add linearly, as they do in Newtonian mechanics. Suppose a projectile is fired from a moving space ship, and both velocities are in the same direction. Suppose that each velocity is $\frac{3}{4}$ the velocity of light. By Newtonian mechanics, we would get $1\frac{1}{2}c$ for the velocity of the projectile, which is forbidden by item 5 above. Einstein showed that the sum of the two velocities, each less than c, will always be observed to be less than c.

We can summarize the three results of motion: lengths are foreshortened; times go slower; masses get heavier. The other five results can be summarized: matter can be transformed into energy, and vice versa; moving bodies are restricted to velocities below 300,000 km/s; magnetic forces are really electrokinetic forces; we live in a four-dimensional universe; and velocities add nonlinearly in space-time.

These amazing consequences of Einstein's two innocent-sounding postulates—one of which he borrowed from Newton, and the other asserting that c is an absolute constant—have all been demonstrated experimentally, many times and in many ways. There are other consequences, too. One of the best known is called the clock or twin paradox. It is a dramatic example of a relativistic effect if spaceships travel at a significant fraction of the speed of light.

Twin Paradox

Archie and Betty are twins. Archie and his clock stay on earth, but Betty, his twin sister, takes her clock on a space trip to planet P. When Betty returns to earth, she is younger than Archie, and her clock reads an earlier time than Archie's clock. *Note*: Betty does not return younger than when she blasted off—we haven't found the fountain of youth yet—but she has aged less than her brother, and her clock shows it. Space travel would probably be funded more generously by elderly politicians if it made travelers younger.

If the planet P is very distant, and Betty travels at nearly the speed of light, the time differentials could be very large. Betty might return, after only a few months by her own time measurement, to find that her great-grandchildren have died of old age. A Rip Van Winkle situation.

The twin paradox seems to violate our intuitive concept of symmetry between Archie's and Betty's frames of reference. Why can't we regard Archie as moving away from Betty, with the spaceship as at rest and the earth moving away from it? Then, it would seem, Archie should be younger than Betty when the twins are reunited, again in the same frame of reference. There are several ways to analyze the problem, each giving the answer that Betty will indeed be younger.

First of all, Betty is the only one who *changes* frame of reference. Archie stays in the same frame throughout, but Betty does not. This would seem to imply that Betty's *accelerations* might be the key to the different clock readings, but this is not so, as can be shown by a third sibling, a triplet, Charlene. Charlene has the same blastoff acceleration as Betty, and the same turnaround accelerations when she reaches her destination. The only difference between Betty's schedule and Charlene's is that Charlene cruises for a longer time (although at the same velocity as Betty), and goes to a more distant planet. Charlene will return younger than Betty. Clearly, how much younger Charlene will be than Betty can only depend on the duration and magnitude of their cruising velocities, relative (as always) to the velocity of light.

Another asymmetry between Archie and Betty concerns the distance to planet P. In whose frame of reference is this distance measured? Measured distances vary with the velocity of the observer. Betty will find that the distance to planet P is not as far as Archie had determined it from measurements on earth, which is nearly at rest with respect to planet P. For Betty, distances in the earth's frame of reference are foreshortened in the direction of her motion.

Another way of seeing the asymmetry, and also why Charlene will age less than Betty, is that when the turnaround occurs, only the astronaut involved, that is, Betty or Charlene, sees the changes caused by the turnaround immediately. The others must wait to learn of the turnaround until the signals reach them. That is, we set up the experiment so that each clock gives off signals, which are radiated to the siblings at velocity c, at fixed time intervals, throughout the trips. Archie, Betty, and Charlene are thus in constant communication throughout the experiment, and they record the time intervals between the signals they receive. (Of course, their own signals, that is, the signals they *send*, show no variation, being separated by equal time intervals throughout the trip.) The time intervals between the signals will be received closer together by Archie when Betty and Charlene are on their homeward trips, compared to the time intervals when they are outward bound. The frequency of the time signals is affected by a Doppler shift.[1]

Stepping back from the details, we can state the problem in terms of two (or three, if Charlene is included) trajectories in space-time. Betty's trajectory intersects Archie's twice, at the beginning and end of her trip. However, that does not mean that the distances (in space-time) between the intersection points must be equal. If we really want to set up symmetrical systems, then Archie must take a trip too. Archie could blast off at the same time as Betty, and go in a different direction. Alternatively, Archie could join Betty on planet P, blasting off at a later time. However, if Archie blasts off after Betty, and goes faster, catching up with her, Archie will be the one who is younger when they return.

Symmetries are important in physics, but they are tricky. As Einstein said, ''God is subtle. . . .'' For example, the image reflected by a mirror is reversed left to right, but not upside down. Why the apparent asymmetry? Does the mirror know one direction from another, or feel gravity? (Hint: There are three dimensions to the image, not two, and you see the back side of the image, along one dimension.)

As Einstein got to the end of his special theory of relativity road, one big question loomed up ahead. This was the apparent asymmetry in the universe between *uniform* motion and *accelerated* motion. Only the former seemed to allow light signals to come under the theoretical umbrella of a principle whereby different frames of reference could be treated as merely moving relative to each other. In Chapter 7 we see the road that Einstein traveled to a more general theory of relativity, including *all* motions.

---

[1]Many scientists have published analyses of this fascinating problem. Two of the best, in my opinion, are: G. David Scott, ''On Solutions of the Clock Paradox,'' *Am. J. Physics* 27:580 (1959); and George O. Abell, *Exploration of the Universe*, 4th ed., CBS College Publishing, 1982, pp. 250–252. Each of the twins records the time signals throughout the experiment; the accumulation of discrepancies is discussed.

6

Quantum Theory: New Phenomena, New Principles

The Quantum Road is the longest, most tortuous, and philosophically the bumpiest of the six roads we are traveling from Newton. Quantum theory involves the most radical change in scientific viewpoint since the Renaissance, and the intellectual revolution is not finished. The loss of an absolute space and time, with relativity, was bad enough, but quantum theory undermines a still more basic belief: causality. Relativity theory gives us a new absolute to grasp, an invariant velocity characteristic of the physical universe, c. Quantum theory gives us an absolute that is harder to grasp, an invariant called Planck's constant, h. Planck's constant has the units of energy times time or, equivalently, of momentum times distance; it is called *action*, and we measure its magnitude, but none of these descriptions gives an adequate clue as to its deeper meaning.

Planck's constant seems to have a deeper meaning because it pops up everywhere throughout the new physics. Beginning with Max Planck's solution to the problem of blackbody radiation in 1900, quantum theory has spread so that hardly any field of physics is without its quantum version. And usually, the quantum version is the decisive, "modern" version. From electronics to optics, from chemistry to astronomy, from radioactivity to lasers, to superconductors, transistors, superfluidity, the atomic nucleus—the applications of the quantum point of view are pervasive. Lurking behind this point of view, like a Svengali watching the successes of the quantum Trilby, is a probabilistic, acausal doctrine. The God of the quantum universe plays dice all the time; what is more, he plays by special rules.

Seven quantum experiments, or phenomena, are described in this chapter. Many more could be presented, but these suffice to present an overwhelming case, establishing scientific facts that convincingly support a theory inconsistent with classical physics. In chronological order the seven phenomena are: blackbody radiation (1900), the photoelectric effect (1905), line series in the spectrum of hydrogen light (1913), the Compton effect (1923), the diffraction of particle beams (1927), the Ramsauer effect (1927), and the tunnel effect (1928). As each of these effects was studied, and the quantum theory grew to explain them, the unhappiness of many physicists grew also. The disputes climaxed in what amounted to a declaration of war by Einstein, Podolsky, and Rosen, in a famous paper published in 1935.[1] What is disputed is the sufficiency, or completeness, of the quantum theory's explanations. In the opinion of many physicists today, the dispute has not yet been settled satisfactorily. In their opinion, the interdependence of causality and probability is still not clearly understood.

Each of the seven named experiments has been performed many times; several of them are laboratory exercises in university physics courses. The experimental results are beyond challenge. Also beyond challenge is the insufficiency of Newtonian, Maxwellian, or relativistic theories, even with ingenious stretching, for explaining these results. In Chapter 9 a viewpoint is outlined that is neither Einstein's nor that of Einstein's opponents. But first, let us describe the seven experiments.

Blackbody Radiation

Quantum theory originated in nothing, a hole, an empty hole. A small hole in a large, empty container is a blackbody because any electromagnetic radiation can enter, and almost none of it will be reflected back out. Such a container, or cavity, is called black because it absorbs all colors (wavelengths) without discrimination.

If the cavity is heated, it emits radiation through the hole. It is like a furnace with a small door. Remarkably, it does not matter what color the

---

[1]A. Einstein, B. Podolsky, and N. Rosen, "Can Quantum-Mechanical Description of Physical Reality Be Considered Complete?" Originally published in *Phys. Rev.* 47:777–780, (1935). Reprinted in the anthology, *Quantum Theory and Measurement*, edited by J.A. Wheeler and W.H. Zurek. Princeton University Press. 1983.

inside walls are painted, or whether they are painted at all. Nor does it matter whether the walls are made of metal or ceramic. All parts of the inside walls must be able to "see" each other, that is, have line-of-sight exposure to each other, so that the radiation exchange (emissions, absorptions, reflections, reemissions, reabsorptions, etc.) can take place between all parts of the cavity wall at the same temperature, producing radiation equilibrium in the cavity. The hole should be small compared to the inside wall area, so that radiation bounces around without escaping too easily. These are the only requirements to make a cavity a good blackbody. Most solid bodies, even if they are not cavities, radiate approximately as blackbodies; stars (even though they are not solid), most nonmetallic surfaces, and even human skin are examples.

The hotter the walls of the cavity, the brighter (more intense) the emission. Not only the intensity, but also the color changes with temperature, going from red hot, through orange, to white. Eventually, if the cavity can withstand the temperature, the emitted light becomes bluish. Blue stars are hotter than red stars. The theoretical problem for physicists was to predict, and explain, these intensity and color changes quantitatively. Since the kinds of atoms in the cavity walls did not affect the emissions, the radiation emitted involved a fundamental relationship between electromagnetic radiation and all forms of solid matter. The physicists wanted to predict how much light of each color would be emitted from the cavity hole at each given temperature inside the cavity. A graph of emitted intensity versus wavelength of emitted radiation is shown in Figure 15.

The experimental measurements are straightforward. The light coming out of the cavity is dispersed by a spectrometer, and then a small detector is scanned slowly through the spectrum, recording the intensity at each wavelength. The cavity is held at constant temperature during this measurement. Figure 15 shows several such curves, each drawn for a different cavity temperature.

The big problem was not the measurements, but the derivation, using thermodynamic principles, of an equation that would fit the experimental data curve, that is, predict the intensity of the light measured at each wavelength. Such an equation was derived and named after its discoverers: the Rayleigh-Jeans equation or law. The intensity I of the light is

$$I = \frac{8\pi kT}{\lambda^4},$$

where k is the basic unit of entropy, known as Boltzmann's constant, T is the temperature of the cavity, and λ is the wavelength of the light.

Unfortunately, it is wrong. The wavelength (to the fourth power yet!) is in the denominator, which means that as we measure shorter and shorter wavelengths of the light emitted from the cavity, its intensity keeps on increasing, without limit. But this would mean that as we passed from the visible spectrum on into the ultraviolet, the curve would eventually violate the law of conservation of energy. As can be seen in Figure 15, each curve has a definite peak; it passes through a maximum intensity and then declines. It does not keep on increasing, as predicted by the Rayleigh-Jeans law. The Rayleigh-Jeans calculation for a temperature of 5,800 K is plotted as a broken line; it can be seen that it fits the experimental curve for that temperature fairly well near the red end only. Its rising toward infinity in the ultraviolet is known as the "ultraviolet catastrophe."

Attempts by Rayleigh, Jeans, and others to overcome the ultraviolet catastrophe were unsuccessful. Various derivation methods were tried, but the difficulty is fundamental. The Rayleigh-Jeans law is, in fact, derived properly on the basis of classical thermodynamic principles. The difficulty arises

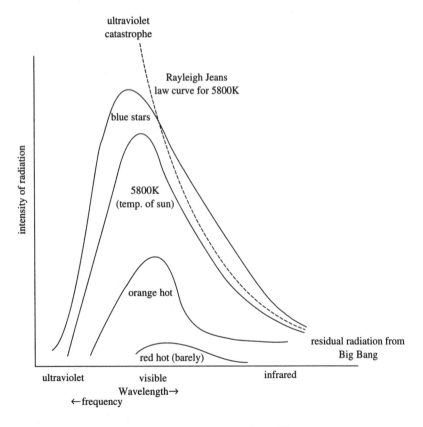

Figure 15 Blackbody radiation curves for different temperatures.

not from the derivations, but from the underlying theory. However, there were some successes on some subsidiary problems. Wilhelm Wien found that the product of the temperature T of the blackbody curve multiplied by the wavelength at which it radiates most intensely λ_{max} is a constant. Wien's law is

$$\lambda_{max} \, T = 2,900$$

when the wavelength is expressed in micrometers and the temperature in Kelvin degrees. This is a highly successful formula; among its applications, it gives the temperature of a star. You just measure the wavelength in the star's spectrum at which its intensity curve peaks, substitute it into Wien's law, and solve for T. Another application: If we want to know the optimum temperature of a blackbody, say an incandescent tungsten filament, for providing illumination for the human eye, we put into Wien's law the wavelength at which the eye is most sensitive (in the green), which is about half a micrometer. If the blackbody is hotter, too much of its radiation would be in the invisible ultraviolet; if the blackbody is cooler, too much of its radiation would be in the invisible infrared. The optimum temperature calculates to be

$$\frac{2,900}{0.5} = 5,800 \text{ K.}$$

This is the temperature of the "surface" of the sun. The significance of this noncoincidence can be expressed as: God created the sunlight to give mankind the best illumination. Or, we can say that the human eye evolved under sunlight conditions substantially the same as exist today. (The fit to human vision is even better; at twilight, as daylight dims and becomes bluer, our color sensitivity shifts toward the blue [Purkinje phenomenon]).

The more curious reader may want to calculate, using Wien's law, in what part of the spectrum our bodies radiate most intensely. Our bodies (regardless of skin color) radiate approximately as blackbodies, and the result is used in instruments looking for breast cancer, and for seeing people in the dark. Hint: Substitute body temperature, which is 310 K.

Another success in investigating blackbody radiation was achieved by Stefan and Boltzmann, who found that the *total amount* of energy radiated by a blackbody, per unit time, per unit surface area, is proportional to the fourth power of the blackbody temperature. Since we can measure how much radiation a star is emitting, this law enables us to calculate its area, and thus the size of the star, which can be presumed to be approximately spherical.

It remained for Max Planck, however, to hit the grand slam home run, and solve the problem. The four runs were as follows:

1. Planck's law, when graphed, fit the experimental data.
2. It eliminated the ultraviolet catastrophe.

3. It gave Wien's law as a corollary.
4. It gave the Stefan-Boltzmann law as a corollary.

Planck's law, or equation, is

$$I = \frac{8\pi ch}{\lambda^5(e^{\frac{ch}{\lambda kT}} - 1)}$$

where c is the velocity of light and e is the mathematical constant 2.71828....;
e frequently appears in probability calculations. Planck's law involves the
statistical behavior of the very large number of atoms radiating in the cavity
walls; h is Planck's constant. The value of h is 6.6×10^{-34} joule seconds;
this is a very, very small quantity, which is why the quantum jumps, or
discontinuities, are not noticeable. Planck's equation is graphed in Figure 15.

The revolutionary character of Planck's equation centers around h, because it implies a definite, finite lower limit to the gradations of energy
emitted by separate atoms. In classical physics, it was always assumed that
the difference in energy radiated by two atoms in equilibrium with each
other could be as small as we please. Planck's law provides the lower limit
to that difference. Thus, continuous gradations are replaced by stepwise,
discrete, changes. The physical universe not only appears grainy, but the
grain size is a fundamental constant of nature, popping up everywhere.

Planck introduced his hypothesis of steps or, as we say, quantized variations, as an ad hoc expedient to avoid the Rayleigh-Jeans law. One can
follow the derivation of the Rayleigh-Jeans law almost to the end, and then
introduce Planck's hypothesis and get Planck's law. If one does *not* introduce Planck's hypothesis, but allows the energy differences to be arbitrarily
small, approaching zero in the limit, one gets the Rayleigh-Jeans law with
its ultraviolet catastrophe. Planck introduced his hypothesis to repair what
appeared to be a flaw in classical thermodynamics; however, the repair did
not strengthen the structure, but brought the whole house down.

One might have thought there would have been jubilation among physicists over Planck's "home run," that is, the great success of his equation.
But Planck's solution violated known physical laws. So there was a general
feeling that, although Planck had found the correct equation, a different
derivation and interpretation was needed.

Photoelectric Effect

The state of euphoria over the success of Planck's equation lasted only five
years. In 1905, three earthshaking papers were published, each written by

the same unknown physicist, who was working as a patent examiner in Bern, Switzerland. It was probably the last time that major breakthroughs were made by anyone working alone, without connection to a university, institute, or research laboratory, One of the three papers established the reality of molecules; no one had ever seen a molecule, and many physicists doubted they existed. This was a paper concerned with explaining Brownian motion, which could be observed through a microscope; the paper explained Brownian motion as due to collisions of water molecules with barely visible motes in liquids. A second paper was the famous paper on relativity. And the third paper, which Einstein called his most revolutionary, was on the photoelectric effect, which is the emission of electrons from certain metal surfaces when those surfaces are irradiated with light.

The photoelectric effect was puzzling. When weak red light is shone on a cesium surface, a few slow electrons are emitted. If the red light is made brighter, more electrons are given off, but they still travel slowly (for electrons). If dim blue light is used, there are few electrons, but they have a high velocity. If the blue light is made more intense, more electrons are given off, but their velocities remain the same. If infrared light is used, no electrons are emitted at all, even if the infrared beam is intense. These facts do not make sense in terms of the wave theory of light, as it had been developed by Fresnel and by Maxwell.

Still worse for the wave theory, Einstein calculated that at least several minutes would be needed before enough electromagnetic energy would be absorbed (stored) by the cesium surface so that a typical cesium atom could expel an electron with the observed velocity, or kinetic energy. Only a relatively few cesium atoms in the surface emit electrons. So what happens to the light that is absorbed by atoms that do *not* emit? The classical picture has the light waves being absorbed all over the surface, but to get enough energy stored by the average cesium atom would take considerable time, Einstein figured. However, the photoelectric effect starts in about a nanosecond. The question is: How does the energy get so concentrated as to affect only a few cesium atoms, when the incident light seems to be spread out evenly?

Einstein overcame these difficulties by adopting Planck's hypothesis. There is no connection between blackbody radiation and the photoelectric effect, so using Planck's questionable quantum hypothesis to successfully explain yet another phenomenon placed Planck's hypothesis on much firmer ground. Einstein assumed that the incident light was quantized, behaving as if it were made up of particles, which he called photons. Thus Newton's corpuscular model of light was reintroduced. Einstein used the equation

$$E = hf,$$

where E is the energy of the photon, f is the frequency of the light, and h is Planck's constant. In words, the energy of a photon is proportional to its frequency (and inversely proportional to its wavelength).

The experimental facts of the photoelectric effect then fall into place. Only those cesium atoms that are hit by photons emit electrons. Blue photons have more energy, so the electrons that blue photons cause to be emitted have more energy—velocity—than those from red photons. From the classical viewpoint, emitted electron velocity ought to increase with increasing incident light intensity, regardless of the light's frequency.

But what, then, becomes of the wave theory of light? Niels Bohr supplied an answer, called *complementarity*. This is a principle according to which the incident light is in limbo, incompletely defined, until it encounters some sort of detector. Then the light becomes either corpuscular or wave, depending on the type of interaction it has with the apparatus. In an interferometer, light behaves as wave trains; in a photoelectric cell, it behaves as photons.

During the next three decades, 1905 to 1935, the classical viewpoint took many more defeats. Quantum ideas were extended to cover more and more phenomena, where classical theory proved inadequate. Quantum theory, as will become clear along Quantum Road, is an extremely versatile, powerful, and self-consistent set of ideas, principles, and rules. If one judged a theory solely by the success of its predictions, the quantum theory would be almost beyond criticism. But what shall we think of an honest and candid racetrack bettor who always picks the winning horse, but cannot satisfactorily explain how he does it?

Line Series in the Spectra of Radiating Atoms

Neon signs emit light when electrons (currents) are pushed through the neon gas in the tube. The spectrum of neon light is complicated, but if we substitute hydrogen for neon in the tube, we get a simpler spectrum, consisting of only a few lines (in the spectroscope), spaced according to an empirical numerical formula, called the Ritz formula:

$$f = R \left(\frac{1}{2^2} - \frac{1}{n_2^2} \right);$$

f is the frequency of the spectrum line, n_2 stands for an integer greater than 2, that is, 3, 4, 5, etc., and R is a constant called the Rydberg constant. A spectrum line appears for each integer value of n_2, starting with 3; the whole series is known as the Balmer series. If n_2 is 3, we get the famous hydrogen alpha line, which is red, giving the hydrogen light its pink color. When n_2 is 4, we get a blue green line, (the beta line), and so on.

If the first term is changed to 1 (instead of ¼) we get another series of lines, called the Lyman series; this time the spectrum lines lie in the ultraviolet because the difference between the two terms is greater, meaning a greater frequency of the light. The Ritz formula is really

$$f = R \left(\frac{1}{n_1^2} - \frac{1}{n_2^2} \right),$$

where n_2 has to be an integer greater than n_1. There are other line series, in the infrared, for $n_1 = 3$ and $n_1 = 4$. For each series, n_2 starts at $n_1 + 1$, with the lines in a series getting closer together as n_2 increases.

This means that the states of the hydrogen atom are quantized, so that emission of light can occur only at certain frequencies. But why? An atom is envisaged as a miniature solar system, with a heavy nucleus; in hydrogen, the nucleus is a single proton, which weighs 1,836 times as much as an electron. In the solar system, according to Newton's laws, a planet can orbit at *any* distance from the sun, so long as it goes in an ellipse and at the right speed (see Kepler's laws, in Chapter 10). But in atoms, even the simplest, which is hydrogen, the requirements for stability are different, being stable only at certain energy levels. When an atom loses energy, that is, emits a burst of light, it changes from one stable energy level to another. It is like jumping out of a hotel window; you can jump from the twelfth or thirteenth story, but not the 12.2 story. If you hit the ground level ($n_1 = 1$) (fall as far as possible), that is like the Lyman series. But if you land on an abutment one story up ($n_1 = 2$), that is like the Balmer series.

There are quantum selection rules: not all jumps are permissible. You can jump only from certain floors to certain other floors. Comparison can be made with the marriage taboos in a primitive tribe. A girl from the wolf totem can marry a boy from the eagle totem, but she may not have anything to do with a boy in the beaver totem. However, atoms sometimes behave like Romeo and Juliet; the wrong couple gets together. These couplings in atomic spectra are called forbidden lines; they occur mainly in astronomical spectra, from sources in which the physical conditions are very different from those in earth laboratories.

There are quite a few selection rules, some dealing with other quantum numbers besides n_1 and n_2. These relate to other properties of the atomic state, such as electron spin. One of the strongest taboos is the Pauli exclusion principle, which states that two electrons in an atom can never have the same set of quantum numbers. The periodic table of chemical elements can be explained as a consequence of the exclusion principle. We can get only so many electrons into a single orbit before quantum numbers duplicate. So the Pauli rule requires that if there is one more electron than can be accommodated in an atomic orbit, that electron must be in a separate orbit.

Niels Bohr *derived* the Ritz formula in 1913, obtaining a value for the Rydberg constant which agreed with experiment. He assumed $E = hf$, as Planck and Einstein had. He also suspended the validity of a classical law, assuming that an electron traveling in a circular or elliptical orbit would not radiate any energy. AC circuits and TV antennas prove the opposite.

Physicists were forced to allow classical laws of electric fields to be violated within the atom. But Bohr's model was neat. The circumference of the electron orbit was quantized because there had to be an integer number of wavelengths in it, like the number of teeth in a gear. However, the Bohr model did not work out when extended to heavier atoms. The whole picture had to be abandoned in favor of a more abstract, mathematical approach, which yielded no nice mechanical picture, such as planetary orbits. In the more abstract version of quantum theory, initiated by Heisenberg and by Schrödinger, the distinction between waves and particles is blurred. In the photoelectric effect, light waves behave like particles (photons). Another such case is the Compton effect, discussed in the following section. Then we will describe three converse phenomena, wherein particles behave like waves.

Compton Effect

When light is scattered by small spheres, Maxwell's equations are the starting point of the calculations. But when the light is x-rays (very short wavelengths), and the spheres are electrons, Arthur Compton found that the scattered x-rays have a lower frequency than they had before they were scattered. This is *not* in accordance with Maxwell's equations. If we consider the x-rays to be photons, with their energy given by $E = hf$ as in the photoelectric effect, lower frequency means lower energy. The energy lost by the x-ray photons shows up in the increased energy of the impacted electrons. So energy is conserved in the collisions, and so is momentum, but why do the x-rays behave like balls, losing energy at collisions? The behavior of light as if it were made of particles thus is not limited to the photoelectric effect.

Diffraction of Particle Beams

We next consider particle phenomena, which follow patterns that in classical physics are associated only with waves. This wave-particle symmetry, or dualism, was predicted by Louis de Broglie. De Broglie reasoned on the basis of relativity theory, where waves are represented by vectors in space-time. De Broglie applied the same space-time formalism to moving particles, and predicted that, by analogy, each particle of mass m and velocity v would show behavior as if it had a wavelength λ of

$$\lambda = \frac{h}{mv}.$$

De Broglie developed this idea in his Ph.D. thesis, submitted to the physics faculty at the Sorbonne, in Paris. The physics faculty faced a dilemma. If they accepted de Broglie's thesis, and it was nonsense, they would lose face. If they rejected the thesis, and it proved to have merit, they would lose face. So they adopted a most unusual procedure: they sent the manuscript to Einstein, who had no connection with the Sorbonne. Einstein was so impressed with de Broglie's idea that he wrote a short introduction. Later, de Broglie won a Nobel Prize for this work.

Experimental verification was right at hand. Davisson and Germer demonstrated that a beam of electrons was diffracted by thin films of metal (which form a sort of diffraction grating), or by crystals. The basic equation which fit the Davisson-Germer measurements was the same as de Broglie's. In words, a moving particle diffracts as if it had a wavelength λ equal to the ratio of Planck's constant, h, to the momentum mv of the particle. The implication is that atomic particles (not only electrons) cannot be thought of as like tiny ball bearings; when moving, they have a structure, or else they have associated waves of some kind.

The diffraction of particle beams recalls the discussion on Wave Road concerning the distinction between diffraction and interference. When waves pass through a narrow aperture, or pass a sharp edge, they curl (diffract) around the obstruction and enter the region of the geometric shadow, where they would not enter if they traveled in perfectly straight lines. We can see this with water waves as they pass a jetty and deviate slightly into the quiet, protected waters behind it. At a given point on the shore, portions of the diffracted wave front are superimposed. They arrive in different phases because they have traveled different distances from the jetty. The superposition of different portions of the same wave front is what we have called diffraction. Interference is also a superposition, but of entirely separate waves. This is the case with the double slit experiment, where there is a separate system of waves from each slit.

By this definition, a diffraction grating, which has many slits closely spaced, is really exhibiting interference. Such a fine distinction between diffraction and interference is not usually of importance, but, in the case of particle beams, there is a real ambiguity. Diffraction here implies that each particle encounters only a single aperture, but is somehow guided by an associated wave of some kind, which is diffracted. Interference, on the other hand, implies that each particle is accompanied by a relatively large wave, which passes through a number of adjacent apertures at the same time. We shall not attempt to establish which of these scenarios better fits the experimental facts.

De Broglie's work inspired Werner Heisenberg to develop a more abstract quantum theory than Bohr's, which is today known as the old or early quantum theory. Also inspired was Erwin Schrödinger, who developed a different abstract quantum theory. Heisenberg used matrices, based on observable state quantities; Schrödinger used a wave (really a diffusion) equation, based on operators representing the momentum and position of a particle. At first, the two formalisms seem quite different, but Schrödinger soon showed that they are completely equivalent. Although other formalisms were developed later, especially one by Dirac and another by Feynman, Heisenberg's and Schrödinger's are still the most frequently used. With these tools, quantum theory has solved problems in every branch of physics.

All four of these formalisms are highly abstract in that they do not provide any *picture* of physical process, any model of what actually happens. They are acausal, giving only the probability of changes of atomic states. The method is thus like that of calculating dice odds, with no account taken of what *causes* the dice to come up the way they do. The reliance is on mathematical rules that lead to calculation of the probability of this or that atomic state being observed. Bohr's picture of the atom as a miniature solar system was scrapped; physicists conjecture that perhaps electrons form a cloud, like bees, around atomic nuclei, with exact location of individual electrons not well defined. What Bohr interpreted as orbits are merely the *most probable* situations, or states. Perhaps each electron only comes together as a locatable particle (like a ghost materializing) when conditions are right and/or a measurement is made. This idea finds support in nuclear physics, because electrons (beta rays) are emitted from certain atomic nuclei, but these nuclei do not contain electrons! They are born on the run.

Ramsauer Effect

If the velocity of electrons in a beam is slowly varied, a certain velocity may be found at which the electrons pass through certain gases without being scattered by the gas atoms, or only being scattered slightly. This strange "transparency" of the gas is not explainable on a classical basis; classically, we would regard the situation like that of throwing basketballs into a forest of trees, and calculating how far a ball is likely to go before colliding with a tree, its mean free path. One uses for the calculation the diameter of the basketballs, and the diameter and spacing of the trees. But quantum theory treats this disappearance of scattering, called the Ramsauer effect, in terms of moving electrons having an effective wavelength, as indicated by de Broglie's equation. If the waves are the right length, they interfere destructively with what would be the scattered waves, so that the latter do not appear.

If we retain the terminology of particle physics, we can speak of the scatter cross section of the electrons and of the gas atoms, that is, the diameters of basketballs and trees. Then we would say that the scatter cross section decreases with the velocity of the electrons. This cannot be attributed to the charge on the electron, because a similar phenomenon is observed with neutrons going through graphite; uncharged particles have the same de Broglie wavelengths. The size of an object (particle), and also its position, is not as simple or unambiguous as it was in Newton's system.

The Ramsauer effect is described above as a resonance effect, due to the matching of the wavelengths associated with the electrons to the size of the gas atoms. However, we can also describe the Ramsauer effect as electrons tunneling through the field comprised of gas atoms.

Tunnel Effect

We have already met the tunnel effect in two guises. The first time was on the Wave Road where it was called total internal reflection. Waves encountering a barrier (reflecting surface) "try" to pass through, and if the barrier is not too strong or too thick, some of the waves will be transmitted through it. The critical factor is the nearness of a favorable medium for the waves on the far side of the barrier.

The second time we met the tunnel effect was as a way of looking at the Ramsauer effect. The paradoxical aspect of the tunnel effect is that some particles are able to "tunnel" through potential barriers that, by classical calculation, should stop them cold.

We can compare the situation with crashing test cars into a brick wall. What are the chances that a car will penetrate all the way through the wall, and emerge on the other side? Suppose we calculate that the chances are one in a hundred. In quantum mechanics, the conclusion would be that if we crash many cars into the wall, about one in a hundred will be found on the other side. We actually "crash" electrons through barriers in a device called a tunnel diode, which is useful in electronics.

The tunnel effect is the model for radioactivity, whereby nuclear particles and energy are emitted through a field that acts as a barrier. The field is comprised of nuclear forces, holding the nucleus together, and the emission process is regarded as purely happenstance, without a triggering cause. We describe the probability of emission in terms of its reciprocal, the half-life of the nucleus; the longer the time period in which half the atoms emit their particle, the lower the probability of finding such an emission in any given measuring period. The concept is similar to that of a mortality table.

However, we can avoid specifically invoking probability if we regard the tunnel effect as the analog, for particles, of total internal reflection for waves. The particle tentatively penetrates into the barrier, and if it reaches more favorable terrain, it continues (is emitted). If not, it returns to the nucleus. The intermediate case of partial reflection and partial transmission recalls Newton's conjecture on Wave Road of the splitting of the incident beam between "fits" of reflection and "fits" of transmission. Newton's formulation is closer to that of quantum theory than is the tentative penetration model, because quantum theory focuses on the individual case, a single particle at a time. Since a single particle cannot be divided into two parts (one transmitted, one reflected), we are forced to a probability approach, like heads and tails for a coin toss. One of the main themes of quantum theory is wave-particle dualism. In the case of the tunnel effect, this dualism can be preserved in either of two ways: (a) by a corpuscular model of light (photons), or (b) by a wave model of particles (de Broglie waves).

Some Quantum Principles

The seven major quantum effects just described illustrate the major principles of quantum theory. Each of them also deals a telling blow at classical physics, which fails to explain any of the seven; worse, in most cases, classical theories give an explanation, but one that is manifestly wrong. Whatever criticisms can be made of quantum theory, and some will be pointed out in Chapter 9, quantum theory is necessary, as we understand the physical universe today.

Let us summarize what we have found. Unless we assume quantization of atomic emissions of electromagnetic energy from a blackbody, the ultraviolet catastrophe is unavoidable. A similar quantization is essential to explain the series of lines found in atomic spectra. The key to quantum theory definitely includes Planck's constant of quantization, h. If h were to get smaller, and go to zero, that is, if the universe were not quantized, then spectral lines would be infinitely close—continuous. Classical behavior is thus a limiting case, when h goes to zero. This principle is called the *Correspondence principle* of quantum theory. We met a similar principle near the end of Relativity Road, when we observed that as the ratio of the velocity of a frame of reference to the velocity of light goes to zero, the relativity equations reduce to Newtonian form.

Wave-particle dualism is another essential part of the quantum theory. We saw two cases of waves behaving like particles: the photoelectric and Compton effects. We saw three cases of particles exhibiting wavelike properties: the diffraction of particle beams, and the Ramsauer and tunnel effects. What determines when waves and particles will act out of character? It seems to depend on what we use to make the observation. In the case of a light beam, we can use an interferometer, and see wave behavior, or a photocell, and see photon behavior. We can even let the fringes formed in an interferometer fall onto a photodiode, which will exhibit the photoelectric effect. To take another case, we observe classical scattering of particles by other particles under usual conditions, but if there is a match of de Broglie wavelengths, we observe the Ramsauer effect. Bohr called the dependence of either wave or particle behavior on the conditions of observation, the *Complementarity principle*. An exact formulation of this principle is still being debated among physicists.

A third major characteristic of quantum theory is its dependence on probability. We have not discussed this in detail because it involves more mathematics than we wish to require of the reader. Worse, there are special statistical rules; Big Jule's dice are not as absurd as they seem. It turns out that there are two different kinds of indistinguishable dice, one following what are called Bose-Einstein statistics and the other following Fermi-Dirac statistics. Every subatomic particle behaves according to one, but never both, of these two types of statistics. The reader will encounter references to bosons and fermions in the literature, referring to the two "tribes" of particles. The rules are very useful and necessary, but their deeper significance is not well understood. Relying on probability, and arbitrary rules into the bargain, means that quantum descriptions of phenomena are basically acausal. The mortality table lets us predict how many people will die next year, but does not give a clue as to what they will die from.

Quantum theory shows atomic events to be related to each other more closely than classical physics would have led us to suspect. The best known of these hidden connections is Heisenberg's celebrated *uncertainty principle*, which shows a hidden interdependence between certain pairs of measurements. That is, when we make an accurate measurement, it can turn out that we have imposed limits on how closely we can measure something else, apparently unrelated. The product of the pair of measurements must have the dimensions of Planck's constant, of action. Momentum times distance, energy times time, and angular momentum times spin angle are the three pairs of measurements to which the uncertainty principle applies. It does not apply to every pair of measurements, just these three, and so should not be loaded with too much philosophical baggage. Heisenberg's principle states that if we measure the momentum of a particle within a small range of uncertainty (probable error), then the simultaneous measurement of the particle's location (position) must have an uncertainty such that the product of the two uncertainties is at least h.

A rough analogy, not to be pushed too far, can be seen when we shift a car into low gear going up a steep hill. We increase the car's climbing ability, but at the same time, necessarily decrease its speed. A trade-off.

The uncertainty principle was not accepted easily. Einstein tried for a while to find exceptions, that is, violations of Heisenberg's relation, and some of them are very interesting. But they were ultimately as unsuccessful as Maxwell's demon was in violating the Second Law of thermodynamics. Einstein had to admit that he failed. He remained, however, unconvinced of the completeness of quantum mechanical models. A prime example of the sort of situation that divides physicists into those who are satisfied with the completeness of quantum theory and those convinced that there must be something else, is provided by Schrödinger's cat.

Schrödinger's Cat

Planck's constant is very small, so the paradoxes of quantum theory are mainly confined to atomic processes. However, we can have a physical system wherein a *macroscopic* event is *triggered* by an atomic event. Schrödinger described a

> quite ridiculous case(s). A cat is penned up in a steel chamber, along with the following diabolical device . . . in a Geiger counter there is a tiny bit of radioactive substance, so small, that perhaps in the course of one hour one of the atoms decays, but also, with equal probability, perhaps none; if it happens, the (Geiger) counter tube discharges and through a relay releases a hammer which shatters a small flask of hydrocyanic acid. If one has left this entire system to itself for an hour, one would say that the cat still lives if meanwhile no atom has decayed. The first atomic decay would have poisoned (the cat). The psi function of the entire system would express this by having in it the living and the dead cat . . . mixed or smeared out in equal parts.

The "psi function" is the wave function in Schrödinger's equation; this function indicates the probabilities for each of the two outcomes. Until the outcome is established by observation, the equation expresses a "mixed or smeared" outcome.

The way in which quantum theory deals with the case of Schrödinger's cat has inspired much discussion and controversy. Schrödinger's cat is the quantum version of the old conundrum: if a tree falls in the forest, with no one to hear it, does it make a sound?'' There are at least five ways of interpreting the situation:

1. Newtonian (common sense) view. Objective reality is completely separate from the observer. The cat is either alive or dead, and our knowledge as to which is true is simply irrelevant. As Einstein put it, it is ''entirely unacceptable'' to deny independent reality ''as such to things which are spatially separated from each other.''[2] Since the observer is spatially separated from the cat, each has independent reality.

2. Quantum (complementarity) view. The state of the cat is represented by its wave function. This function is undetermined until the observation is made. One state then becomes well defined at the expense of the other, because the system interacts with a measuring apparatus, i.e. the observer. (Perhaps the cat itself can qualify as a ''measuring apparatus,'' or maybe the Geiger counter can qualify.) The situation recalls the baseball umpire defining a strike: ''They ain't nothin' till I call 'em.''

3. Many Worlds view. The emission of the radioactive particle splits *reality* into two branches, or worlds. When we look into the box, we split our reality into two more branches (The cat is alive in one, and dead in the other.) All these possibilities exist; there are ''skillions'' of realities. The terms of the wave function do not collapse or disappear, as in the conventional quantum view, above.

4. Statistical view. The wave function applies to a large number of similarly prepared systems, in each of which the cat is either alive or dead. As time increases, the number of systems in which the cat is dead increases (because of the increasing probability of its death), while the number in which it is alive decreases. When we make the observation, that precipitates the definitive result, replacing the probabilistic situation with a single situation. This is similar to the situation of tossing a penny.

5. Reality (existence) is a matter of degree. We are concerned, not only with the cat, but also with our observation of the cat; as we increase our information, the cat's degree of existence (or death), relative to us, increases. The wave function has two terms, one representing each of the two probabilities (decay or no decay). The paradox disappears if we abandon the presumption that we are in possession of all possible information about the atomic decay (radioactive emission of a particle). We are simply shooting dice.

The common sense view assumes that the cat is there to the same degree whether we look at it or not. Information counts for something in the phys-

[2]*Albert Einstein: Philosopher-Scientist*, edited by P.A. Schilpp. Library of Living Philosophers, Tudor Publishing Co. New York, 1951. P. 85.

ical universe, not just subjectively. The complementarity interpretation errs in the opposite direction, attributing physical reality only to the final state. The Many Worlds view has been supported by some physicists, but it would seem to deprive physical reality of all but a subjective meaning, as seems to be the case in some Oriental philosophies. The fourth (statistical) view fails to take into account that quantum theory often gives definitive, even if not exact, answers. Statistical results are not substitutes for results in individual cases, but are additional to specific results. The fifth view, the reader is warned, is my own minority opinion.

The dispute over Schrödinger's cat may sound like a parlor game for quibbling physicists. But that cat is in a combat zone, a philosophical no-man's land concerning the nature of physical reality. *First*, it sweeps away the fence between macroscopic everyday reality and the domain of atomic events, with its strange quantum rules. Physicists have been trying to locate, or relocate, a fence between the two for a long time.[3] However, some quantum phenomena are directly observable on the macroscopic scale, using samples as big as a cat. Superconductivity, superfluidity, and the specific heat of solids at low temperatures are well-known examples.

Second, Schrödinger's cat presents in dramatic fashion what physical reality is like if it is given only a probabilistic meaning. The Schrödinger equation is a field equation, and the field variable is *probability*. On Field Road we saw fields of force—gravitational, electric, magnetic—and considered fields of other physical quantities, such as temperature and color. But in Schrödinger's equation the field variations consist of greater and lesser *odds*. Worse, these variations can be periodic—wavelike—and can interfere constructively and destructively.

Third, Schrödinger's cat raises the question of quantum theory's completeness, which is the target of Einstein's criticism. Quantum theorists, especially John Von Neumann, have offered proofs for believing that all physically possible information is contained in Schrödinger's equation. Hidden variables, that is, other causal factors, are specifically prohibited, as inconsistent with a very powerful and successful theory, a theory, moreover, which has no rivals.

More will be said about each of these three "cat issues" in Chapter 9, where their implications will be focused toward a single viewpoint. We have traveled quickly on Quantum Road; no mention has been made of the different effects which, in alphabetical order, have been named for Hall, Josephson, Lamb, Mossbauer, Stark, and Zeeman. But enough has been indicated to suggest the magnitude of the intellectual revolution wrought by quantum theory. The question still remains of whether or not scientists should be

---

[3]W.H. Zurek, "Decoherence and the Transition from Quantum to Classical." *Physics Today* Oct. 1991. Letters commenting on this article are printed in the April 1993 issue.

satisfied with a theory that parallels the data but does not explain it. Systematic organization of experimental data, and even successful prediction of such data, does not yet constitute a complete scientific theory. Einstein was the most prominent of those who expressed such points of view. Einstein was a major role model for scientists, because with respect to the limitations of his own restricted theory of relativity, discussed in a previous chapter, he remained dissatisfied because of its restrictions. He worked, virtually alone, for 11 years, to generalize his theory of relativity.

7
General Relativity: Gravity as Field Distortions

The special theory of relativity is restricted, as Newton's first law is, to uniform motion in a straight line. But the question thus far unanswered is: Does the principle of relativity (that the motion of a body is relative to the observer's motion) apply to *accelerated* motions, especially rotations, like that of the earth? Isn't it possible to tell absolutely, whether or not a body is spinning, without considering its relationship to any other body? The centrifugal force on a spinning object, especially obvious in a liquid, is attributed to the rotary motion with respect to empty space. This was Newton's point of view, and it was not questioned until Einstein wondered if the relativity principle might not be applicable to all motions.

In Newton's time the character of empty space was discussed in different terms. Gottfried Leibniz, a contemporary of Newton, objected to the implication, in Newton's *Principia*, that empty space could be endowed with *physical* properties. Space, for Leibniz, was the turf of mathematicians and philosophers like himself and Euclid. Whether empty space could be an ether, supporting light wave propagation, or gravitational forces, was an open question. To show that his viewpoint was correct, Newton proposed the following gedanken experiment, that is, an experiment we can discuss

logically without actually performing it. The experiment is known as Newton's pail.

A pail of water is suspended from a rope, and the rope is twisted, rotating
the pail by hand. When we release the pail, the rope will untwist, and the
pail will rotate in the reverse direction. Consider the five phases of events:

1. Before we release the pail, there is no relative motion between the pail,
   the water in the pail, or the earth. The surface of the water in the pail is
   flat (horizontal).
2. We release the pail, and it begins to rotate. The pail walls are smooth, so
   there is little friction between the walls and the water. Thus, when the
   pail rotates, the water remains stationary, having inertia (Newton's first
   law). The water sits there, while the walls of the pail slip past it. Since
   the water is not moving, its surface remains flat, even though there is
   relative motion between the water and the pail.
3. There is a little friction, however, between the water and the walls of the
   pail. So gradually, the frictional force on the water makes it partake of
   the pail's rotation. There is no longer relative motion between the water
   and the pail, but the surface of the water is concave (paraboloidal) due to
   centrifugal force.
4. The rope unwinds, the pail comes to rest, but the spinning water has
   inertia, and keeps on moving. There is again relative motion between the
   water and the pail, but this time the surface of the water is paraboloidal.
5. Everything comes to rest, and returns to the initial conditions, No. 1
   above.

Why does the surface of the water become paraboloidal? Is it because
the water moves relative to the earth, to the pail, or to the ether? If we were
to do the same experiment on a spaceship far out in space, we might expect
to get similar results, but we would not, because we would not be in a strong
gravitational field. The water seems to exhibit absolute motion.

Straight-line motion, at uniform velocity, seems to obey the relativity
principle, but accelerated motions do not. Other examples come to mind. If
we are in a car that lurches forward, or slams on the brakes, or turns a sharp
corner (each of which is an acceleration), we passengers can tell, without
looking out the window at any other frame of reference.

This implies that the physical universe is divided into two parts, in only
one of which the principle of relativity is valid. Physicists dislike such du-
alisms. Special relativity was fashioned to eliminate a threatened dualism
between mechanical signals (ballistics) and light signals (optics). Einstein
hoped to do the same for accelerated motion, that is, to remove the privileged
status of rotating frames of reference, so that the principle of relativity would
apply to *all* frames.

At first, one would expect that Newtons's laws, plus special relativity, plus field equations analogous to Maxwell's, might do the trick. No such luck. After 11 years, Einstein succeeded in creating a field theory of gravity quite different in concept from Newton and from Maxwell.

Before we plunge ahead, consider the Newtonian view again. Centrifugal force is real, not an illusion of a rotating observer, as anyone who has worked with a centrifuge can testify. The spinning earth is flattened by 13 miles at each of its poles, and bulges at the equator. These distortions of the earth's sphere can be seen from Mars as well as on the earth, both on and off the rotating frame of reference.

If the earth were to rotate once per *hour*, instead of once per day, would people in the tropics tend to fly off? You bet! The formula for acceleration of a rotating body is the velocity squared, divided by the radius (distance from the axis of rotation):

$$a = \frac{v^2}{r}.$$

The circumference of the earth is easy to remember, because the meter was defined as a ten-millionth of a quadrant of a great circle of the earth through Paris. That makes the circumference 4×10^7 meters, and the radius is the circumference divided by 2π. The acceleration due to gravity is 9.8 m/s^2. This not only applies to bodies in free fall toward the earth, but by symmetry, also applies to a body *leaving* the earth. When we substitute the numbers for velocity (one rotation in 24 hours, expressed in meters per second) and the radius, we find that 9.8 m/s^2 is reached when the earth turns only 17 times faster than it does now. Another way to look at it is that at one rotation per hour the earth's equator is going 25,000 miles per hour; we send our spaceships away from earth at about 18,000 miles per hour. Would it help people in the tropics to stay on earth if they wore lead in their shoes? No, remember a heavy body and a light body are accelerated the same by gravitational pull.

If people were to fly off the earth, would they fly off tangentially, or radially (straight up)? We discussed this question in Chapter 1 (Newton's laws for astronauts), and concluded that the answer, due to Newton's first law, was: tangentially, if you were not on the rotating body, but radially if you were. So we would see our friends fly straight up (radially), but Martians watching the fun would see them fly tangentially. So, there is some relativity to centrifugal force after all. If our friends fly off tangentially, it is just their inertia, (continuing in the same direction), instead of their changing directions as the earth rotates. So we can "transform away" the centrifugal force, by changing our viewing position to a reference frame that does not share the rotation.

Another example of a "spurious" force we can observe is the Coriolis force. The air in the tropics is heated, and rises, and spreads. Consider the air that spreads north. It cools after a while, and descends to near the earth's surface. But that surface, being quite a bit north of the equator, is not rotating as fast as the equator is. The air, however, still has the greater rotational velocity it had near the equator. So the air is moving faster than the surface it approaches. This is the source of the prevailing westerly winds. But is there a "real" force pushing these winds? The answer, again, depends on your frame of reference.

If centrifugal force can be transformed away by a suitable change of reference frame, how about doing the same thing for gravity? In the discussion of weightlessness in Chapter 1, we saw that in free fall, gravity seemed to disappear. One way of describing weightlessness is the apparent absence of gravitational force on a mass. A man in a free-fall elevator would find that a shot he is holding does not fall on his feet if he stops holding it.

How do we know that force is being exerted on a body? Because the body accelerates. To what do we attribute the force? The fact that acceleration is produced. This circular argument disturbed Einstein. There had even been a serious attempt, by Hertz, to eliminate force as a primary variable in mechanics. Along similar lines, Einstein was impressed with Mach's attempts to relate unaccelerated motion, not "relatively to space, but relatively to the center of all the other masses in the universe," that is, to the stars. Einstein objected to saying that empty space had "a physical effect, but (was) not itself influenced by physical conditions." He concluded, "The cause [of gravity] must therefore lie *outside* this system."

Einstein saw that the starting point for a unified theory had to relate inertia to gravity, for which no one, including Newton, had found a clear-cut causal explanation. Newton had, however, observed a strange coincidence: The mass of a body as measured by its inertia (reluctance to change its velocity), was the same as its mass as measured by gravitational attraction for, and by, another mass. The equivalence of inertial mass and gravitational mass was assumed in Chapter 1 to show that a heavy and a light body fall together. Einstein elevated Newton's coincidence into the principle of equivalence: The inertial and gravitational mass of a body are two manifestations of the same thing, and must be identically equal. Gravity and inertia are different manifestations of the same basic phenomenon. The quantitative equivalence of gravitational and inertial mass has been measured to about one part in 10^9.

Einstein gave a gedanken experiment to offset, although it does not refute, Newton's pail. Einstein uses an elevator suspended in space. The man in the elevator cannot look out, but he measures falling bodies inside the elevator. He may think that the bodies fall because the earth is below him and the earth's gravity is pulling them down; in that case, there is a gravitational force on the bodies, and the elevator is suspended on a sky hook above the earth. Alternatively, if the elevator is far out in space, so that no mass is exerting gravitational pull on it, the elevator could be towed by a rocket accelerating it upward, and in this case the falling bodies would appear to be falling because of inertia. Bodies would fall downward inside the elevator either way, from gravity or from inertia. The principle of equivalence forbids any effect or measurement that would allow the man to distinguish between the two cases.

Unfortunately for the simplicity of the theory, the man in the elevator can try three methods, or experiments, to distinguish between gravity and inertia. We can get an insight into the theory, without getting involved with the difficult mathematics, by considering these three experiments.

First, if the man compares the path of a body falling near one side of his elevator with the path of a body falling near the opposite side, he will find that the paths converge slightly, instead of being exactly parallel, if earth gravity is pulling the bodies downward. The two paths will converge toward the center of the earth, which is spherical. However, if the elevator is being towed upward by a rocket, the paths of the two falling bodies will be strictly parallel to each other.

Second, if he measures the acceleration of the bodies falling in his elevator, the man will find that their acceleration is slightly greater near the floor than near the ceiling, if earth gravity is doing the pulling. This is because the floor is nearer the earth than the ceiling is, and hence the gravitational attraction on the falling bodies is a little greater. However, if the elevator is being towed upward by a rocket, the acceleration will be the same at every point within the elevator.

Third, if he shines a light beam horizontally across the elevator, gravity (earth pull) should have no effect on the light path, if light has no mass. However, if the elevator is being towed upward by a rocket, the light beam will hit the far side of the elevator a little nearer to the floor. This recalls the case, in discussing stellar aberration on Special Relativity Road, when a cannonball was fired through a moving ship; however, the ship was moving at uniform velocity, and the elevator is accelerating. The two effects, although apparently similar, are quite different.

A lesser man than Einstein would have thrown in the towel at this point. After he had successfully formulated a mathematical theory that eliminates these three differences, he wrote,

> The possibility of explaining the numerical equality of inertia and gravitation by the unity of their nature gives to the general theory of relativity, according to my conviction, such a superiority over the conceptions of classical mechanics, that all the difficulties encountered in development must be considered as small in comparison.

Einstein did not build the theory by concentrating on the three experiments, above. Instead, he followed a mathematical route using a four-dimensional, non-Euclidean geometry. We shall not follow the mathematics, but pick up the theory further down the road, pointing out some of its features.

The first two experiments show that the general theory must be infinitesimal, that is, must be applied to negligibly small "elevators." As we reduce the dimensions inside the elevator toward zero, the differences noted as to convergence of paths and rates of falling acceleration also reduce to zero. Thus the general theory of relativity is a "local" field theory, where the properties of the four-dimensional space vary from point to point.

The third experiment pointed to an extension of the special theory of relativity, wherein light *does* have mass, according to the relation $E = mc^2$. The general theory extends this to include the effects on light caused by gravitational fields. In the case of the elevator experiment, the light beam would bend as it propagated across the elevator, the same amount as it would bend due to a certain amount of acceleration. The field equations of general relativity thus have to indicate how strong a gravitational field matches how great an acceleration. In other words, the field equations are built on the principle of equivalence. The bending of light from a star, as the starlight passes near the sun, has been measured by photographing the position of a star that appears near the sun (as seen from earth) during a total eclipse of the sun. The star's position is then measured at another time of year, when the star is seen far from the sun. The bending is very small, less than 2 seconds of arc, which is barely beyond the probable error of the measurements. Thus this experimental evidence is not very powerful.

A more observable effect predicted by the theory is the perturbation of planetary orbits. The elliptical orbits of the planets, predicted by Newton's theory, do not remain quite fixed. The ellipse as a whole creeps slowly around the sun so that the major axis of the ellipse changes its direction. This effect is greatest in the case of Mercury, and it had been observed before the relativity theory came along to explain it.

A third observable effect predicted by the theory is an increase in the wavelength of light emitted by a star, due to the gravitational effect of the star on the emitting atoms. This gravitational red shift is quite different from the Doppler red shift, discussed in Chapter 5. In the case of the sun, the gravitational red shift is very small, but the effect has been observed in light coming from some very dense stars. The effect has been observed on earth, too, despite the earth's weak gravitational field, by use of a quantum "trick" involving the Mossbauer effect.

The three observed effects are small, but the effect of the theory on our concept of space and time is not small. Space-time is conceived as being distorted, curved, rather than laid out neatly for Euclidean triangles and spheres. When a body "falls," that is, responds to gravitational attraction between itself and another mass, it follows the shortest available path. If the gravitational attraction is relatively weak, as it is in all our earthbound experiences, then the body falls very nearly in a straight line, because a straight line is then the shortest distance between the two centers of gravity. *Light* propagates in straight lines, too, in space which has only moderate gravitational fields. Shortest paths are called geodesics, or geodesic paths. There seems to be a general characteristic of physical systems to follow geodesics, although in some cases they are not minimum paths but maximum paths. We have already seen one case of geodesy, that of Fermat's principle for light going through different optical media. General relativity provides another; other cases are mentioned in Chapter 9.

It would have been nice if gravity were just a special inertial force, as magnetism turned out to be just a special electrical force in the special relativity theory. But gravity exists between masses where there is no relative motion; magnetism does not exist where there is no relative motion between electric charges. The key to gravity lies not in motion, but in distortions of space-time. In other words, gravity and inertia are equivalent, but they are not identical. They have an underlying unity, and are equal in magnitude, like four quarters and a dollar bill.

One of the most startling predictions of the general relativity theory concerns completely collapsed stars. If light is so bent near an extremely massive star that the light turns completely around, the light could not leave the star at all. Or maybe the gravitational red shift is so great that the frequency of the light goes to zero: It has no frequency, and hence no existence. Most astronomers believe that there are collapsed stars of stupendous densities that are capable of this; they are called black holes. They are black, but they are not holes. In fact, they are really the opposite of holes: intensely dense masses, tremendously powerful celestial whirlpools, sucking in (by gravity) the gaseous matter from any star which approaches. We even think we have located a few black holes, from the effects we observe near them.

Special relativity showed that distances, and time intervals, are dependent on velocity, a correlation that was not suspected earlier. In addition, mass was shown to be equivalent to energy, by the famous equation $E = mc^2$. General relativity goes further and shows that mass determines the geometry of the space surrounding it, another surprising correlation. Thus general relativity completely rejects Leibniz's claim that space is physically empty. Space is a physical field, although not the nice Euclidean field that Newton implied with his pail experiment.

Einstein hoped to extend the field concept to include all physical phenomena, especially electrical. Such a field was called, in anticipation, the *Unified Field Theory*. But the field equations of Maxwell are very different from the field equations of general relativity. No one has yet succeeded in ''marrying'' them. Many physicists, marking Einstein's lack of success, doubt if such a theory is possible.

In the next chapter we take a few peeks down some of the roads physics has traveled since quantum mechanics and relativity brought about their revolutions in scientific thinking. We will not describe most of the work that has been done, because there is so much of it, and because much of it requires advanced technical and mathematical preparation. However, we can point out some of the hidden, and surprising, connections that have turned up.

8

A Look Down Further Roads

Nuclear Physics

Ancient Greek philosophers hypothesized that the subdivision of matter into smaller and smaller pieces had to have an end; there had to be, they thought, an unsplittable particle that formed the basic building bricks of matter. This idea was taken over by Renaissance scientists and carried into modern science. However, the development of modern chemistry required that the atom, as the basic brick was called, itself have a structure. At first, the atom was considered to be composed of two kinds of particles having opposite electric charge. The negative particles were called electrons, and the positive particles were called protons. Although the charges are equal in intensity, the proton has a mass 1,836 times greater than the electron.

At the end of the nineteenth century, Joseph Thomson, in England, worked on a model of the atom in which the protons and electrons were distributed somewhat like nuts and raisins in a cake. Ernest Rutherford, also in England, bombarded some heavy atoms with alpha particles, and observed the scattering of the alpha particles. In 1911, Rutherford described a model of the atom in which most of its mass was concentrated in a nucleus of protons, with the electrons buzzing around, somewhat like planets around a star. The picture became clearer in 1932 when James Chadwick, also in England, identified another type of particle, the neutron. Neutrons are uncharged, and somehow help to hold the protons together (since like charges repel each other) in the nucleus. All atomic nuclei contain some neutrons, except hydrogen, which has only one proton and so needs no "glue."

When neutrons were discovered, the general assumption was that each neutron was a fusion of a proton and an electron. However, when two fundamental particles fuse, there is always a loss of mass; the fused particle weighs less than the sum of the weights of its separate constituents. This is a relativistic effect. But the neutron mass is slightly *more* than the sum of the proton mass and the electron mass. So something more complicated is involved. The present theory is that protons and neutrons are composed of more elementary particles, called *quarks*.

The study of the nucleus and its forces has been a long road, and the end is not yet reached. One nuclear model is called the liquid drop because the nucleus sometimes seems to minimize its own size, as drops of liquid do, due to surface tension. Another nuclear model is called the cloudy crystal ball because the nucleus is somewhat transparent to certain radiations, yet also scatters them. One of the achievements of quantum mechanics was its explanation of why the electrons don't fall into the nucleus of the atom, collapsing the atom. Nuclear physics has been preoccupied with a reverse problem, explaining why the nuclear protons don't fly apart.

Forces and Particles

To explain why the nucleus holds together, physicists hypothesized a new force of nature. We already are familiar with gravity, which Newton dealt with and Einstein refined. The electric force is a second kind of force. The nuclear force has to be stronger than the electric force, in order to hold the protons together despite their electrostatic repulsion. However, on the macroscopic scale—at greater distances—protons *do* move apart. So the assumption is that the nuclear force does not decrease inversely as the square of the distance, but much more rapidly than that. Thus this force, called the strong nuclear force, dominates when the protons are very close together, as in the nucleus, but the force becomes negligible at larger distances.

A similar problem, with a similar solution, concerns the spontaneous emission of particles and energy from certain atomic nuclei. This is called radioactivity, and the main things emitted are shown as emanations in Chart 3. This time physicists hypothesized a weak nuclear force. Like the strong nuclear force, it falls off very rapidly with distance.

Thus, we have four basic forces: gravitational, the weak nuclear force, electrical, and the strong nuclear force. Gravity is so much the weakest of these four that it does not appear within atoms at all. The electrical force between two electrons is about 10^{40} times stronger than their gravitational attraction for each other. To give an idea of this enormous number: an ant pushing or pulling the whole earth would only have to be stronger than the typical ant by about 6×10^{28}, a number about 100 billion times less than 10^{40}. These figures give an idea of how closely the plus and minus charges in ordinary matter must balance each other (and thus cancel each other out) in order that we can experience gravity in everyday life.

Physicists respect Occam's razor, so many are unhappy with having four basic forces. A lot of effort has gone into finding hidden correlations between these forces, that is, unified field theories. There has been some success, particularly in marrying the weak nuclear force to the electric force. Electric and magnetic forces are described by Maxwell's equations, but these equations have continuous variables, and so do not indicate any graininess in electric fields. To introduce some graininess, or fine structure, into the fields, the theory was modified. The quantum theory provided the model, and the new theory is called QED, quantum electrodynamics; it introduces Planck's constant into the field equations.

Besides the electron, the proton, and the neutron, physicists have found many new subatomic particles, many of which live less than a second before disintegrating into still smaller particles and into gamma rays. The old idea of the unsplittable atom is as obsolete as Ptolemy's geocentric model of the solar system.

Chart 2, a Particle Chart, shows some of the properties of just a few of the simpler particles. There are well over 100 different species in the subatomic zoo. They come in different weight classes, like boxers; the lightest are leptons, the middleweights are mesons, and the heavyweights are baryons. Every type of particle has its ghostly twin, an antiparticle: The proton has an antiproton; the electron's twin is called the positron. Even the uncharged particles, like the neutron and the neutrinos, have their antiparticles. An exception is the photon, which does not have a distinguishable antiphoton. To save the rule, physicists say that the photon is its own antiparticle; alternatively, one could doubt that photons exist as separate particles.

The particle zoo has many weird species. Neutrinos have zero rest mass, but they are never at rest, so they have a relativistic mass. In a solid semiconductor, there may be a hole (an empty apartment) in the lattice of the solid. An electron from ''next door'' can slip into this hole, leaving the former position of the electron empty. Thus the hole is displaced by one position, in the opposite direction from the electron motion. The process can continue, with the electron current going one way, and the hole current going the other way. The hole is thus like a particle, a virtual particle.

Chart 2 Properties of Simple Particles

Name	Picture	Symbol	Charge	Approximate Mass (Atomic Units)	Other Name	Atomic Species	Penetrating Power	Abundance	Natural Radioactive Emanation
Electron	⊖	e	−1	$\frac{1}{1836}$	Beta ray β cathode ray	None	Absorbed by air	Abundant everywhere; easily obtained from metals	Yes
Proton	⊕	p	+1	1	Hydrogen nucleus	$_1H^1$	Absorbed by air	Abundant in nuclei	No
Neutron	Ⓝ	n	None	1	None	None	Penetrating; absorbed by water and certain nuclei	Abundant in nuclei	Yes
Gamma ray	∿	γ	None	0	Hard x-rays (photons)	None	Very penetrating; absorbed by lead or other high density materials	Can be generated by beta rays	Yes
Alpha particle	(⁺⁺nn)	α	+2	4	Helium nucleus	$_2He^4$	Stopped by paper	Found in helium gas	Yes
Deuteron	(⁺n)	D	+1	2	Deuterium nucleus	$_1H^2$	Easily stopped	0.015% of the hydrogen in seawater	No
Triton	(⁺nn)	T	+1	3	Tritium nucleus	$_1H^3$	Easily stopped; but emits β particles	Synthetic; half-life is 12 years	No

Nuclear Energy

One of the interesting roads building on the discovery of nuclear energy is the pursuit of controlled fusion of atomic nuclei, along the lines of what happens in the sun. This is a vastly more efficient means of getting energy from fuel than burning it, and somewhat more efficient than splitting atomic nuclei, as we did in fission bombs at the birth of the atomic age. The distinguishing characteristic of all forms of nuclear (also called atomic) energy is that the energy derives from a transformation of (loss of) mass, according to the relativity equation, $E = mc^2$. The food we eat and the fuels we burn involve chemical reactions, especially oxidation, an altogether different, and far less efficient, process. A lump of coal, if its mass were transformed entirely into energy, could supply more than enough energy to fly an airplane from New York to Paris. Moreover, it wouldn't have to be burnable coal; it could be a plain old rock.

One might think, as one student of mine did, that this formula equating energy and mass could be used to figure out how many calories of energy you should eat, or rather not eat, in order to equal the mass (weight) you want to lose on your diet. It is easy to compute (just remember that 1 food calorie equals 1,000 of the physicists' calories), but the result is disappointing. If you were to convert the entire mass of the food you eat into energy (by nuclear energy transformation), then one hamburger, or whatever, is all you need to eat. You have ingested enough energy to carry you through more than a billion days. How many days do you expect to diet?

As shown in Chart 3, there are four forms of nuclear energy conversion:

1. *Fission.* When heavy atomic nuclei are bombarded with neutrons, the nuclei break up (fission) into smaller pieces, with some mass disappearing. If you drop a glass on the floor, then sweep up the pieces and weigh them, you might find that some mass is missing because some shards went under the bureau. In the nuclear case, however, some mass really *is* missing; it has been transformed into energy, such as the blast of the A (atomic) bomb, and the heat in a nuclear reactor.

2. *Radioactive emissions.* Unstable atomic nuclei, instead of fissioning, may emit a single particle or pulse of electromagnetic energy. Sometimes this occurs spontaneously, and sometimes we induce (stimulate) it, similar to the way we induce fission. Different things are emitted; the four most important, and dangerous, are indicated in Chart 3.

3. *Annihilation/creation.* When a particle meets its antiparticle, for example, an electron meets a positron, the two collide and annihilate each other. At the scene of the tragedy, a gamma ray appears, and radiates away. This process is reversible; under certain conditions, not too well understood, a

gamma ray will suddenly transform into a particle and its antiparticle, that is, the particles are created in pairs.

4. *Fusion.* This is the reverse of fission. Particles or nuclei are forced to fuse into one mass. There is a loss of mass involved, which appears as energy; that is, the fused mass is not quite as heavy as its constituent pieces were when they were single. The created energy is the energy of the sun and the stars, and of the H (hydrogen) bomb. Fission generally involves heavy nuclei, especially uranium and plutonium. Fusion generally involves light nuclei, especially hydrogen and helium. The sun is mostly hydrogen and helium.

We have been trying to produce controlled fusion. We know the theory, and it works in the stars and in the H bomb, but we would like to fuse hydrogen without destroying the county. The theory, based on the equation $E = mc^2$ seems straightforward, but the engineering difficulties have stymied all the nations that have participated in the effort for more than 30 years. The basic trouble is that we cannot reproduce the temperatures, pressures, and concentrations that are maintained inside the sun and the stars. The payoff will be much better than superbombs; it will probably give us a permanent solution to the energy crisis. We could get more energy than is good for us just using seawater.

Fusion does not have toxic end products, the way fission does. Our

Chart 3   Nuclear Energy Chart

$E = mc^2$

fission
- A bomb $_{92}U^{235}$ (Hiroshima); $_{94}Pu^{239}$ (Nagasaki)
- breeder reactors—produce plutonium $_{94}Pu^{239}$ from ordinary uranium $_{92}U^{235}$
- burner reactors—generate steam to generate electricity

emanation from nuclei (radioactivity)
- spontaneous
- induced

alpha rays	(helium nuclei)
beta rays	(electrons)
gamma rays	(electromagnetic waves)
neutrons	(induce radioactivity and fission)

annihilation/creation matter and antimatter $\longleftrightarrow$ electromagnetic pulse (waves) (gamma rays)

fusion
- H bomb Bikini test 1954
- sun and stars—helium is the end product
- controlled fusion not yet achieved

"best" fissionable material is plutonium, probably the most toxic substance known to man. So far. There are no plutonium mines; it is all made artificially. Plutonium is an alpha particle emitter. Alpha particles are easily stopped; you don't need lead underwear, as you do with gamma rays and neutrons. If there is an explosion in a reactor, plutonium dust might be thrown up into the air. If there is a meltdown, plutonium vapor might go into the air. If a microscopic amount of plutonium gets through your gas mask, it can lodge in your lung, still giving off alpha particles. You might, if you have a macabre sense of humor, think of getting a free chest x-ray, from the inside out, but the alpha particles cannot penetrate the chest wall. Instead, they will be absorbed by lung tissue, in all probability causing a cancer at the site. The plutonium in a single breeder reactor, if spread around the world from a reactor accident or terrorist attack, could kill every man, woman, and child in the whole world. So let's not talk any more about that.

PCT

Particles moving in a beam, such as an electric current, exhibit many of the properties of waves. One of the differences between waves and particles relates to polarization. Transverse waves, like light waves and water waves, can be made to oscillate in specific directions perpendicular to their direction of propagation. Particles have an analogous property; they can spin. If each particle in a beam spins about an axis through itself, as the earth and other planets do, and all the spin axes are parallel to each other, this gives the particle beam some of the characteristics of polarized light.

A well-known textbook[1] says, "Spin is often pictured by assuming the particle to be a fast-spinning top. However, for any acceptable radius of the particle, the velocity at the surface of the particle then exceeds the velocity of light, and the picture therefore is not really tenable. In addition, even particles with zero rest mass, such as the photon and the neutrino, possess a spin. The existence of spin has to be accepted as a fact." The last statement illustrates a subject that has fired the controversies; spin is not a fact, but a convenient fiction for an incompletely understood type of particle polarization.

Parity is the technical term for direction of spin rotation. A clock can go clockwise or counterclockwise. Parity also refers to left-hand and right-hand screw threads, and to x–y–z coordinate axes, which can be like a right-handed glove or a left-handed glove. The equipartition principle, discussed

---

[1]H. Frauenfelder and E. M. Henley, *Subatomic Physics*, Prentice-Hall, Inc. Englewood Cliffs, N.J., 1974. p. 70.

in Chapter 4, indicates that nature should have no preference as to parity; left-handed and right-handed systems, spins, DNA molecules, and so on should be equally likely and equally prevalent. If we had to explain which is our right hand and which is our left to ETs on a distant planet, we would be hard put unless we could point to some spinning celestial body that we and they could both see.

That's what we used to think. But in 1957, physicists discovered that in one type of subatomic process, those involving the weak nuclear force, parity is *not* conserved. Nature shows a definite preference. The experimental facts are well established, and the appropriate Nobel prizes awarded. Interpretation is something else again. Some physicists say that it "has to be accepted as a fact." But some of us "insist on dragging metaphysical considerations into the laboratory." We say that there must be a hidden correlation between this violation of equipartition and something else. The cards are marked somehow. The most likely suspects are *charge conjugation* (transformation of matter into antimatter, and vice versa) and *time reversal* (yes, time running backward). With parity as the original suspect, the interaction of the three is known as the PCT puzzle.

What is involved is a matter of symmetry. P and C and T may not be symmetrical when separated, but in some combinations, the symmetry should emerge. A left-handed batter has an advantage in baseball because the asymmetrical choice was made, probably by a left-handed inventor of the game, to run around the bases counterclockwise. Symmetry can be tricky, as we saw in Chapter 5 in the discussion of the twin paradox.

Physicists are strongly prejudiced in favor of symmetry. We want to treat absorption processes as symmetrical (reversed) emission processes, attraction as symmetrical to repulsion, positive and negative charges as similar cases of the same basic effect, and so on. This prejudice is partly based on Occam's razor, but goes deeper. Symmetrical processes represent the equipartition principle, with the implication that physical causes reveal themselves through violation of equipartition or symmetry. We say, if the penny falls heads half the time and tails half the time, it just happens, no special causative agent need be hypothesized. But if you keep tossing heads, one after the other, I am going to start suspecting hidden variables.

Coming back to the PCT puzzle, there is an interesting feature. You cannot make a smooth, continuous transition from a left-handed system to a right-handed system. There is no close-fitting glove intermediate between the left-handed one and the right-handed one. The transition is quantized. So is charge conjugation, and so is time reversal. Feynman pointed out that C and T are correlated, that an antiparticle can be treated mathematically as a regular particle that is traveling backward in time.

The arrow of time, which in our experience seems to run always in the same direction, poses a question: why the asymmetry? Some physicists trace it to the expanding universe, some to the Second Law of thermodynamics, some to both. Maybe antiparticles *are* moving backward in time, and we will find that our subjective feelings about past and future are inaccurate. Time will tell.

"Anybody Who's Not Bothered by Bell's Theorem Has to Have Rocks in His Head"

The forward areas of any discipline usually include different interpretations of what is going on. We have touched on a few vortices of confusion, or of controversy, in particular, Schrödinger's cat, Many Worlds, special statistical rules for particles, and the PCT puzzle. In the next chapter we will discuss a favorite gedanken experiment with which quantum theorists demonstrate the need for abandoning causal explanations, namely, Young's double slit experiment used for single quanta. But perhaps the greatest furor centers on a paradox known as Bell's theorem.

In a famous paper, published in 1935, Einstein, Podolsky, and Rosen argued that the quantum theory was incomplete. They proposed a gedanken experiment (known as the EPR experiment) to show that objects, especially quantum objects, possess a physical reality independent of whether they are observed, measured, or disturbed in any way. This is a fundamental matter of principle on which quantum and classical physics differ. As modified by David Bohm, (Chapter 22 of his *Quantum Theory*) the EPR experiment consists of a molecule containing two atoms, with their spins in opposite directions, so that the molecule has zero total angular momentum. The two atoms are separated, without disturbing their spins. If we measure the spin of one of the atoms, we know, without disturbing the other, which direction that other spin has. With classical objects, such as a penny, this experiment is as though we split the penny's thickness in half so that one half is the heads side and the other is the tails side, and mail each half to a different person. A recipient, after looking at only his half of the penny, knows whether the other person has heads or tails.

Quantum theory, however, counters this experiment when, instead of using heads and tails, we use two measurable variables that are related by a quantum relation, such as the Heisenberg uncertainty principle relating momentum and position, or the quantum rule on paired spins, whereby if one spin is up, the other must be down. According to the uncertainty principle, if one of the paired quantities is measured very accurately, the other, no matter how far away it is, cannot simultaneously exist very accurately. So measuring one would seem to have an effect on the other, even when there does not seem to be any way, or time period, for a signal to be propagated between them.

The EPR paper was duly answered, by Bohr, Bohm, and others, but no one, apparently, was convinced, on one side or the other. What each side showed, essentially, is that its view had logical consistency within its own accepted principles, a sort of "you look at it your way, and I look at it in my way" standoff. Einstein and Bohr died with the philosophical dispute between them still unresolved, as it is to this day for many physicists.

However, the controversy again came to a boil after John S. Bell published his proof in 1964, showing that, according to quantum mechanics, the EPR experiment would not work out, that the correlation of the two particles would follow quantum, not classical, statistics. The apparent implication was that there was a "spooky action at a distance." (This is Einstein's phrase, although Einstein died before Bell's theorem was discovered. Many physicists would give a lot to know what Einstein would have said.)

Bell gave a general calculation for an experiment modeled on EPR, but involving quantum theory's restrictions, such as that the spins of two "married" particles, separated but otherwise undisturbed, must remain correlated. Bell showed that the probability of observing matching spins of the two separated particles was different from the probability implied by EPR; that is, there is an inequality that should be observed if we try the experiment many times. An analogy, in the nonquantum world of our everyday experience, is if we draw two marbles from a large box containing an equal number of black and white marbles. By classical statistics, the odds for drawing two black marbles are $1/4$; the odds for drawing two white marbles are also $1/4$; the odds for drawing one of each are $1/2$; that is, 50-50. But in quantum statistics, we have something like Big Jule's dice described in Probability Road, and the odds for the three different outcomes are equal: $1/3$, $1/3$, and $1/3$.

Bell's proof stirred things up because it implied, even more sharply than EPR, how the dispute could be put to an experimental test. The experiments have been done. Engineering complications required some modifications, so diehards are still rooting for EPR, but in my opinion, the game is decided.[2] These experiments, it should be noted, deal with statistical correlations. If there *were* a spooky signal between the two separated particles, it might be expected to work every time. The dice are not fixed to win on every throw; they are only slightly loaded. We must look, I think, not for exceptions to the uncertainty principle, nor for mysterious signals, nor for "spooky action at a distance," but for new interpretations, especially for parts of the setup that are not challenged.

A magician's audience is thrown off the track because their attention is focused on the wrong question: "See, the box is absolutely empty; here, examine

[2]D. Greenberger, M. Horne, and A. Zelinger, "Multiparticle Interferometry and the Superposition Principle." *Physics Today*, Aug. 1993.

it for yourself.'' If you are knowledgeable in prestidigitation, you suspect that the crucial clue lies elsewhere, perhaps in the table the box is placed on, or in the magician's quiet hand. With Bell's theorem, my predisposition is to question the presumptions that the two particles are completely separated and independent while they are moving apart. Perhaps we are confronted with a case of nonlocatability of particles. There may be a symmetry between quantum pairing and spatial relations on the one hand, and particle/antiparticle pairing and time directions on the other; the latter was mentioned in discussing the PCT puzzle.

I don't pretend to have a solution to the paradox presented by Bell's theorem, or causal explanations of the experiments which seem to support it. But while that game may be over, the pennant for the whole season hasn't been won or lost yet, as the next chapter is intended to show. Interest is still high; David Mermin wrote a stirring article on Bell's theorem in *Physics Today* (April 1985).[3] The quotation at the head of this section is taken from Mermin's article.

---

[3] N. David Mermin, ''Is the Moon There When Nobody Looks? Reality and the Quantum Theory.'' *Physics Today*, April 1985.

9

Neither Determinism
Nor Indeterminism

The Need for a More Inclusive View

Each of the six roads that physicists traveled after Newton led them to revisions of viewpoint. The discovery of fields and electromagnetic waves required a thorough revision of the concepts of mechanics for physicists reared in the Great Clockwork tradition of Newtonian physics. Relativity theory required radical revisions of our concepts of space, time, mass, and energy. But even these revisions in fundamental concepts are minor compared to the implications of quantum paradoxes, especially Schrödinger's cat, the double slit experiment for single photons, and Bell's theorem. The very concept of physical causation has been brought into question, and physicists have had to decide whether or not to surrender their efforts to truly explain phenomena and to be content instead with merely describing and predicting them. We saw on Probability Road that this choice arose in predicting dice throws and mortality tables, but then we passed off the need to substitute probabilities for experimental details as merely an expedient adopted for lack of information. Quantum theory has raised the stakes, asserting that such information does not exist, even in principle; for many phenomena, there are no causal factors or hidden variables. We are casting dice for our scientific souls. Does God play dice with the universe? That is, do chance happenings underlie all caused processes, or is it vice versa? Or is the truth somewhere in between?

These questions go to the heart of the controversies of the scientific revolution precipitated by modern physics. In this chapter, an attempt is made to indicate a road to more open country. However, the reader is warned that the way is beset with difficulties, for him, for me, and for everyone else who is interested. Physicists disagree on many of these questions, and the prospects are for more disagreements in the future. Science progresses through disagreements.

Every new, radically different, viewpoint raises the question of its relationship to the old viewpoint. What, exactly, has been changed? How can still valid portions of the old viewpoint be expressed in terms of the new? The equations of relativity theory reduce to those of Newtonian mechanics for slow velocities, that is, as the ratio of the body's velocity v to the velocity of light c reduces to zero: $v/c \to 0$. Transient phenomena merge into steady-state conditions as the time becomes longer and longer. Quantum phenomena go over to classical cases if Planck's constant is reduced to zero.

There seem to be four ways of categorizing the relationship between quantum theory and classical physics. Each is laden with difficulties, including that of defining the boundaries between classical and quantum domains.

1. *Duality view*. Accept the duality (bifurcation) of the physical universe, with different laws for the quantum and classical domains. Bohr assumed that the measuring apparatus was classical, and Bohm's textbook, *Quantum Theory*, adopts a similar viewpoint. But where and how does the changeover between microscopic (atomic, quantum universe) and the macroscopic (ordinarily observable) universe occur? The difficulty is highlighted by Schrödinger's poor cat. The seven quantum phenomena described on Quantum Road are all observable macroscopically, as are superconductivity and the superfluidity of helium.

2. *Quantum view (Copenhagen view)*. Embrace the physical unity of the universe (''monotheism'') on the basis of *quantum* theory as primary and fundamental, and show that classical phenomena are simplified, limiting cases. The majority of physicists have this point of view; it is the viewpoint presented in the three quite different respected textbooks each entitled *Quantum Mechanics*, by Paul Dirac, by Leonard Schiff, and by Richard Feynman and Alfred Hibbs. But under what experimental conditions does the transition between classical and quantum behavior take place?

3. *Classical view*. Embrace the physical unity of the universe on the basis

of *classical* (Newtonian and relativistic) theory as primary and fundamental, with quantum theory as a special case, limited to statistical validity. This is the position implied by the 1935 paper of Einstein, Podolsky, and Rosen, mentioned in connection with Bell's theorem (Chapter 8).

4. *Holistic view.* Embrace the physical unity of the universe, with *both* quantum and classical concepts regarded as simplified, limiting cases. This more general or complete viewpoint is adopted here; it does not postulate strict causality (determinism) nor "pure" happenstance (indeterminism), but a continuum of intermediate cases. A mathematical formalism for such a viewpoint has been proposed by David Bohm (who changed his view from No. 1 above, after discussions with Einstein), but has not led very far.

With relativity, we lost the absoluteness of time, distance, mass, and energy, and the comforting certainties of immutable Euclidean space. With quantum theory, we lost the exact Truth, and the right to presume exact values of physical quantities, and we have injured, perhaps fatally, our belief in causality. In quantum theory, observable events obey laws of chance, but the probability of these events occurring is calculated by mathematical methods similar to those we use for events that follow causal laws. In quantum theory, a sort of virtual causality holds between the probabilities of events, but not between the events themselves.

Between them, classical and quantum views subject the criteria of reality to a procrustean bed, either stretching them beyond their range of validity (classical), or lopping off some essentials (hidden variables). In this view, classical physics makes the Great Clockwork too tight, not allowing tolerances. This is unrealistic, but makes possible the solution of many problems. Quantum theory has a different set of special assumptions and restrictions; these, too, are somewhat unrealistic, but also make possible the solution of many problems. The holistic view to be argued in this chapter thus regards both classical and quantum views as incomplete; physical reality, existence, truth, and causality are matters of degree, depending on quantity of information. This involves a modification of the two-valued logic of Aristotle, whereby a proposition must be either true or not true. Between truth and falsehood, and between existence and nonexistence, there stretches a continuum, a field comprised of information.

Vigorous disputes over these questions have continued for more than half a century. Perhaps we ought to suspect from the no-win struggle between the first three viewpoints listed above that it is not resolvable within any of them. That is, the questions are wrongly put, and we should seek a larger, more general, framework. As in the old mechanism versus vitalism dispute, determinism and indeterminism are each somewhat simplistic, based on too narrow a framing of the questions. One-to-one correspondence between cause and effect will not do; the Great Clockwork is broken, and not even

Einstein could wind it up again. On the other hand, quantum theory needs to be interpreted so as to admit quasi-causal explanations of quantum phenomena.

In the following sections of this chapter, three lines of argument are presented. The first is that transient phases of events are systematically excluded by quantum mechanics, which focuses on initial and final states. The second line of argument suggests modification of the classical picture of electromagnetic waves and of atomic particles, leading to quasi-causal interpretations of quantum phenomena. The third line of argument reconciles causal and statistical descriptions of physical processes through a least action (geodesy) principle and an equipartition principle.

"Foolish Questions"

Two examples of neglected transients were given in the discussion of waves. One involved the explanation of low-reflection coatings on glass, and the other involved the paradox of the three polarizers. A third example, given here, involves a simple DC circuit, consisting only of a battery, a switch, and two resistors connected in parallel; it is shown in Figure 16. Fewer electrons will flow through the higher of the two resistors, in accordance with Ohm's law. The interesting question is: when we close the switch and the first electrons reach the branch point, how do they know in what proportion to divide, that is, how many electrons should flow into each of the parallel branches? The current hasn't reached the resistors yet, so what divides it? Either there is a transient (precursor) electric wave reflected back

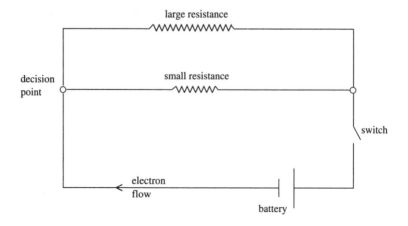

Figure 16 DC parallel circuit.

from each resistor, or else the first electrons just surge through the wires, disregarding Ohm's law; the branch having the greater resistance experiences the greater traffic jam. The backed up electron traffic regulates the current, in the proportions called for by Ohm's law, once steady state is established. There is nothing remarkable in such a picture, although transients are usually overlooked in describing simple DC circuits.

However, there are difficulties in applying this viewpoint to quantum phenomena, such as Schrödinger's cat (see the end of Chapter 6). When the question of neglected transients is raised in quantum cases, the conventional reply is in terms of probability. The probability that Schrödinger's cat is dead increases with time, at a rate dependent on the half-life of the radioactive substance in its cage. This provides us with a time-dependent calculation, but it does not give us a physical process. According to quantum theory, the wave function (as specified by the Schrödinger equation) has two terms, one giving the probability that the cat is alive, the other giving the probability that the cat is dead. Asking—what is the state of the cat *before* we look into the cage?—is like saying that the penny, *before* we toss it, is partly in the heads state and partly in the tails state. In the case of the penny, we put the problem into a mathematical form that ignores the physical factors involved in the toss, but in the case of the cat, we have the additional difficulty that quantum theory *forbids* any hidden variables.

Probability is the branch of mathematics based on the equipartition principle. This covers a wide range of applications. At one end are the applications that disregard causal factors altogether, and consider only random processes leading to possible outcomes (final states), pure chance. Examples are dice, card games, and coin tosses. Then there are applications that take some account of causal factors, but also make some use of random occurrences, or of statistical information. For example, if team A is 2 runs ahead at the end of the 7th inning, what are the chances that team A will win the baseball game? The chances of a particle tunneling through a potential barrier (the tunnel effect) is calculated from a similarly "mixed" viewpoint of causality and randomness. Sometimes the available statistics (as distinct from causal factors) are only partially relevant, for example, what are the odds that the volcano will erupt this week, that it will rain tomorrow, that so-and-so will win the election?

If we base our prediction entirely on causal factors, and disregard random processes as summarized in statistics, we are no longer dealing with probability, even though we may express our confidence in the prediction in probabilistic terms, for example, I'll give you 3 to 1 odds that I am right. Here the probability part of the calculation is at one remove, relating not to making the prediction, but to its probable accuracy or probable error. We are using causal reasoning and calculating, but are not completely certain that we are doing it right.

Quantum mechanics uses probability in a slightly different way. Causal factors are considered, but are not admitted as *processes*. Thus transients can, at most, play a very minor role in quantum theory; the atom is in an initial state, and we compute the probability of finding it in the final state. It is like saying that the penny before we toss it is partly in the heads state and partly in the tails state, and calculating the frequency with which each of these possible future states will be observed. Applying this viewpoint to the problem of the branched DC circuit in Figure 16, we would say that each electron leaving the battery is partly in the lower branch state and partly in the upper branch state. The probability of finding an electron in each is given by Ohm's law. The transient phase, and thus the causal process, is disregarded, as it is in the coin toss.

In the October 1988 issue of *Physics Today*, Herman Feshbach and Victor F. Weisskopf,[1] two well-known theoretical physicists, wrote an article, "Ask a Foolish Question . . ." They assert that "the questions that make sense . . . can be answered. . . . Ask a foolish question and you will get a foolish answer." One of the questions it is foolish to ask is: How long does it take an atom to change its state, when it emits light? A matter of principle is involved because this question has no meaning within quantum theory. Suppose, to take a classical analog, we ask in a dice game: What caused your last throw to come out a 4? The rules of probability cannot deal with such a question; they can only let us calculate the probabilities for each of the outcomes we consider possible. The physical causes affecting a particular throw of the dice, such as how the dice are picked up and held, how hard they are thrown, and on what kind of a surface, are not taken into account. These are "hidden variables." Quantum theory does not simply relegate hidden variables into the closet; it denies that they exist at all. The Schrödinger equation represents steady states, and thus excludes transient processes, hence the assertion that it is foolish to ask the time during which an atom emits. We asked this question at the end of Wave Road, and now ask it again, because in principle we can *measure* this time period with an interferometer. We can also calculate it classically, treating the radiating atom as a kind of oscillator. As pointed out on the Wave Road, a difference in the path length of the two beams in an interferometer leads to the fringe pattern washing out (losing visibility). This path difference is called the coherence length; it indicates, among other things, the length of the wave train. If the path difference between the two arms of the interferometer (Figure 9) is more than half the length of the wave train, the two portions of the wave train being recombined will miss each other—the one having traveled the longer path will arrive too late. If we divide this difference by

---

[1]Herman Feshbach and Victor F. Weisskopf, "Ask a Foolish Question . . ." *Physics Today*, Oct. 1988.

the velocity of light, we get a time, which depends partly on the time during which the atom emitted the wave train. Unfortunately, the measured coherence length is also affected by several other factors, such as the velocity of the atom (Doppler shift) and the electric fields in or near the atom (Stark effect), so experimental ingenuity and theoretical calculations are needed.[2]

The quantum way of looking at the situation is quite different, and does not recognize a process of emission at all. The longer the atom is in the excited state, the more stable the energy level it is in, the greater the uncertainty in the time of emission; therefore, the energy uncertainty (by the uncertainty principle) is very small, and the line is sharp, corresponding, by Fourier analysis, to a long wave train. This could be expected to give a long coherence length, if we are dealing with the wave train from a single atom in the source.

We need not be concerned about the width of the energy level of the ground state because the atom is in that state most of the time. With the residence time very large, the energy uncertainty is negligible, compared to ΔE in the excited state.

We can, a little crudely, compare the two points of view (i.e., holistic vs. quantum) regarding change of atomic state with a family moving from one apartment to another. How long it takes to make the move would seem to have little relation to how long the family has been living in the apartment. The two views are concerned with different time periods: (1) time during which a change is made, and (2) the duration of residence.

One might conjecture that while the atom is in its excited state, something happens to it over time, so that when it emits it is less damped. Then the longer it sits in the excited state, the longer the emitted wave train, and hence the longer the coherence length. (The energy and fundamental frequency, however, must remain the same.) This is an unorthodox conjecture, and I do not urge it, but it illustrates how the uncertainty principle might be interpreted in terms of a hidden variable.

A serious practical difficulty with determining the length of the wave train from the observed coherence length, is that in order to see a fringe pattern we require contributions from a great many wave trains, that is, the emissions from many atoms, firing at random times. Energy levels in atoms are typically complex; we speak of their fine structure, and their hyperfine structure. Each atom's wave train produces its own fringe pattern with its own fringe width variation as the interferometer path difference increases. It is the changing phase relationship between slightly different fringe patterns, which causes the visibility of the superimposed fringes to wash out, thus determining the measured coherence length. Typically, we observe a coher-

[2]H. E. White, *Introduction to Atomic Spectra*. McGraw Hill. New York, 1934. Chap. XXI. See also, M. Born and E. Wolf, *Principles of Optics*, 3rd ed. Pergamon Press. 1965. Sect. 7.5.8.

ence length of a few tenths of millimeters; Michelson found the red line in the spectrum of cadmium to be exceptionally "clean," and got a coherence length of about 200 millimeters. If we take this length as indicating the length of the wave train from each separate emitting cadmium atom, we get a transition time of 0.66 nanoseconds.

This is a pretty iffy calculation, indicating, at best, an order of magnitude of a minimum period of emission. What we really need is to photograph, by time exposure, the interference pattern formed by separate atoms, one at a time, and measure how the pattern visibility on the film fades as we increase the path length difference. (We need one time exposure photograph for each fixed distance setting of the interferometer.) This means working with the emission from one atom at a time. Note that we have said nothing about one quantum of energy; each wave train contains several hundred thousand oscillations (for the red cadmium line, anyway), and this will be a great many quanta. So we need not assume that the interferometer splits a single quantum. Nor have we made any use of the photon hypothesis, another no-splitting-allowed (corpuscular) scheme. However, the discussion shows that the question of transition time has physical meaning whether quantum theory can handle it or not.

This discussion reverses the usual viewpoints. Usually quantum theory falls back on individual cases (e.g., one photon or—not the same thing— one quantum). In this case, the quantum view focuses on the spread (probability distribution) of the energy levels occupied by many different excited atoms, while the view urged here focuses on a single emission at a time.

Another "foolish" question concerns the particular time and direction of a particle "spontaneously" emitted from an atomic nucleus, that is, radioactive emission. According to quantum theory, this is a particular, observable event without any particular cause. It is a case of the tunnel effect, described on Quantum Road, and is dealt with as a probability problem. Just why the particle comes out *now*, not then, and goes *this* way, not that way, is not considered. When we observe a single event, we can legitimately demand "an explanation of when, where and how it is decided what the observer actually perceives."[3]

The question of an alpha particle emitted by an atomic nucleus is dealt with in "The Wave Mechanics of Alpha-Ray Tracks" by N. F. Mott.[4] Mott raises the question of "a definite direction" for the emitted particle, but then substitutes the quite different, and subsidiary, problem of showing that the trajectory must be radial with the source of the particle, that is, along a straight line passing through the atomic nucleus. The demonstration is im-

[3]W. Zurek, "Decoherence and the Transition from Quantum to Classical." *Physics Today*, Oct. 1991.
[4]N. F. Mott, "The Mechanics of Alpha-ray Tracks." *Proc. Ro. Soc.* (London) A126:79–84 (1929). Reprinted in *Quantum Theory and Measurement*. Princeton University Press. 1983.

pressive, but ducks the central question, namely, *which* track the particle
will be observed to take. Here we have an observable phenomenon, a measurable azimuth, and no theoretical tools for handling it. Similarly, we can
measure the time of emission, but we may not even ask of the quantum
theory more than the probability that the emission might occur within a
certain time period. David Bohm's *Quantum Theory*[5] poses the same sort of
questions, but does not answer them. Mortality tables are useful and veri-
fiable, but they do not preclude our asking why so-and-so died.

Causative processes, unlike probabilistic transitions between physical
states, require some continuities in the degrees of freedom (measurable variables) that define the states, usually including the durations and the positions
in space. Dispensing with such continuities may simplify the problem, but
it marks the method as incomplete. Note that causative processes are not,
in this sense, equivalent to determinism. Determinism means: It couldn't
happen in any other way. Causative processes mean: Transient processes are
to be inferred, leading to the observed outcome and its consequences.

If we assume that quantum events, though observable, can originate from
nothing, that there are no *possible* assignable causes, that predictions of
probability are the *complete* story, that there are, *in principle*, no hidden
variables, then we are abdicating scientific searches for explanations. This
is a person's democratic right, of course, but it is a right of religious freedom. Theoreticians, notably J. von Neumann and J. S. Bell, have given
proofs excluding hidden variables of a classical nature from quantum theory,
but no one has demonstrated why the quantum domain must be off-limits
to scientific explanations, with only statistical predictions permitted.

One way that quantum theory could validate its claim to completeness of
information would be to show that what we ordinarily interpret as individual
quantum events, for example, the emission of a single particle from a nucleus, or a single track in a bubble chamber, is illusory, and that the only
event that has real physical meaning is an ensemble, a collection of similar
or related events. To use the coin toss analogy, the theory would say that
we do not observe the results of individual tosses of heads and tails, but
only a population of tosses. Individual events cannot be separately counted,
merely summed. The mortality table is complete if there are no individual
deaths, only collective deaths. However, there is always more to the story
than the best measurements can show. For example, a photograph that
reaches the diffraction limit of the imaging system and the resolution limit
of the photographic emulsion does not contain nearly as much information
about the object photographed as a hologram does. And who is to say that
the hologram has captured everything?

---

[5]D. Bohm, *Quantum Theory*. Prentice-Hall, Inc. New York. 1951. pp. 137–138 and p. 294.

Particles That Behave in Wavelike Fashion

On the Quantum Road, three examples of wavelike behavior by moving particles were described: the tunnel effect, diffraction of particle beams, and the Ramsauer effect. There are others. The wave chart implicitly raises the question of what kind of waves are associated with moving particles. They are sometimes called de Broglie waves, or matter waves, or wave packets. By any name they do not smell as sweet as real waves.

Their wavelength λ varies inversely with their momentum:

$$\lambda = \frac{h}{mv}.$$

This is de Broglie's formula. The slower the particle moves, the longer the wavelength associated with it. The propagation velocity of the wave depends on the velocity of the particle, which can be accelerated or slowed by attractions and repulsions. True waves, such as those of water, sound, and light, have velocities dependent on the properties of the field through which they propagate. In a homogeneous medium, true waves have a constant velocity; they cannot be accelerated.

We can circumvent the idea of waves, and consider instead merely the spread of the particle, or its locatability. We can consider that a particle is spread out in space-time, coalescing into a little sphere under certain conditions. The diffraction angle θ of a beam of particles follows the standard formula for wave diffraction:

$$\sin \theta = \frac{\lambda}{D},$$

where D is the aperture width. Substituting into de Broglie's formula, we eliminate the wavelength and get:

$$\sin \theta = \frac{h}{mvD}.$$

This equation shows that mvD must exceed the magnitude of h, in order for diffraction to be observable; otherwise the value of the sine exceeds unity, which is impossible for real angles. This fits in with Heisenberg's uncertainty principle, whereby the product of the simultaneous uncertainties of the momentum and position of a particle must exceed h. The equation also indicates that particle diffraction will be observable only when mvD exceeds h by a small amount; otherwise, the diffraction angle is unobservably small. Since Planck's constant is a very small quantity, particle diffraction is limited to atomic particles and apertures of molecular dimensions. As for the velocity of the particles, if it is made too slow (to keep mvD small) the beam is lost in thermal vibrations. The diffraction is thus observed only when the value of mvD is neither too large nor too small. This limited range gives it the character of a resonance effect between the particle and the aperture.

The resonance idea also describes the Ramsauer effect, whereby the transmissivity of certain gases for an electron beam increases dramatically when the de Broglie wavelength of the electron beam is just right. An analogy in classical optics is provided by the transparency of crystals to certain wavelengths of light. The spreading of particles so that they have some wave features when they are in motion, as implied by the de Broglie waves, is a radical concept. However, it makes the analogy between the tunnel effect and the phenomenon of total internal reflection in wave optics comprehensible. This seems more credible than the still more radical alternative model of corpuscles of light, as implied by photons.

These speculations should not be taken too seriously until they are supported by other considerations, such as the transformation of electron-positron pairs into gamma waves, and vice versa. These speculations are intended merely to raise what the lawyers call ''a reasonable doubt'' concerning the orthodox quantum explanations of wave-particle dualism.

Waves That Behave in Particlelike Fashion

Two examples of electromagnetic waves behaving like particles (photons) were pointed out on Quantum Road: the photoelectric effect and the Compton effect. A big part of the problem in the photoelectric effect is to explain the very large capture cross section of the metal atoms comprising the photocell surface. This is the reverse of the Ramsauer effect; instead of the capture or collision cross section disappearing, as in the Ramsauer effect, in the photoelectric effect it becomes enormous. Two experimental facts provide clues. The first is the critical dependence of the efficiency of the photoelectric effect on the preparation of the metal surface. Making a photodiode is almost a black art. The critical factors are obviously in the metal surface, not in the incident light beam.

A second significant experimental fact is the large variability of the photoelectric efficiency with the polarization of the incident beam. This too points to a match with the absorbing (and then emitting) atoms. It should be emphasized that the dimensions of the metal atoms emitting photoelectrons are far smaller than the wavelengths of the incident (exciting) light that is absorbed; this means that the usual wave-front model (the Huyghens-Fresnel principle) is inapplicable; a quite different model is needed. We recall a somewhat similar situation along Wave Road, where the spacing of the reflector wire mesh comprising a radiotelescope dish is smaller than the wavelength of the incident radio waves. In the photoelectric effect, we are dealing with the fine-structure of waves. We do not have one to one correspondence between light emitting atoms in the source and electron-emitting atoms in the metal surface. A single photon is not usually the child of a single atom in the source; it is quantized according to energy, not parentage.

What we have, then, is electromagnetic waves being pulled into tiny atoms. The effect resembles a whole swimming pool of water being sucked into relatively tiny drain holes. We are conjecturing that the energy dispersed throughout an electromagnetic wave might be captured by a much smaller absorbing cavity if the cavity is especially tuned.

Expanding this idea, we can conjecture that compression of the energy in an electromagnetic wave might also occur in another way, not by absorption but by piling up on itself, when it encounters certain types of obstacles. This picture could, conceivably, be adapted to the creation of electron-positron pairs by waves in the gamma ray spectral region.

As with the speculation concerning de Broglie waves, in the preceding section, these remarks are meant chiefly to suggest that wave-particle dualism is not a closed subject, but may yield to more convincing explanations than quantum theory has accorded us up till now.

In the Compton effect, as in the photoelectric effect, the wavelength of the incident light is much larger than the particles being targeted. That is, the wavelengths of the x-rays are much larger than the electrons. In consequence we have, not diffraction of the x-rays, but a mechanicallike collision. Despite their difference in size, however, the two colliding parties are fairly well matched in energy and/or in mechanical impedance. Nonetheless, one of the parties to the collision is a wave, and the other is a particle; they are thus unlike, incoherent. Thus the exchange of energy is of intensities, not amplitudes. Mathematically, this means that there is no cross-product term in the superposition. Therefore there is no interference, and the x-rays, although electromagnetic waves, behave like particles.

The Double Slit Experiment Revisited

With slight modification, Young's double slit experiment, described on Wave Road, is useful in presenting the need for quantum theory. Instead of using light of ordinary intensity, however, we use an incident beam so weak (dim) that it provides only one photon going through the apparatus (shown in

Figure 2) at a time. The argument is then made: since the photon cannot split in two, it must go through either one or the other of the slits. How then can we explain interference fringes building up on a photographic film, as the film records the results of successive one-photon transits? Feynman says of this experiment that we have "a phenomenon which is impossible, absolutely impossible, to explain in any classical way, and which has in it the heart of quantum mechanics. In reality, it contains the *only* mystery."[6] Feynman abandons at the outset " 'explaining' how it works. We will just *tell* you how it works." It should be understood that this experiment has never been performed, because it is, today at least, not feasible. We are arguing principles, not data.

First, let us enlarge the issue. (The reader is warned that what follows is a controversial exposition.) We assert:

1. Any two-beam interferometer, such as Michelson's (shown in Figure 9), is equivalent for this gedanken experiment. It makes no difference whether we split the wave train by aperture (slits) or by amplitude (semisilvered mirror).
2. Interference effects are independent of the number of wave trains in the apparatus at the same time. The same explanation should be valid for one, for two, or for a million. Remember Occam's razor.
3. The use of light is not equivalent to the use of electrons in this gedanken experiment. Running one electron at a time, or pairs of electrons, through the apparatus, is a different ball game, as will be argued below.

The experimental fact that a fraction of a photon is never detected, argues not only against their splittability, but also against their existence as separate entities. If we consider that we are using a single wave train emitted by a single atom in the light source, there is no paradox. If both paths (or slits) are open, we get interference; if only one path is open, we do not observe interference. To get interference, we need two (or more) coherent wave trains superimposed on each other. The apparatus creates a pair of coherent wave trains when it splits each wave train; it then recombines them (superposition).

If the wave trains are so short and weak that they contain only a single quantum of energy, we are not dealing with a wave, but with a transient pulse. The rules are different. Probably there will not be any interference, no matter how the apparatus is arranged. This experiment has never been done, as far as I know, either for (1) single-atom emissions, or (2) single quanta, or (3) traceable photons.

What is called the measurement of an individual photon is usually the

---

[6]R. Feynman, *Lectures on Physics*. Addison-Wesley. 1963. Vol. III. 1.1.

detection of ions, as in a Geiger counter; the ions are detected; the photons are inferred. Let us assume, however, that somehow we arrange to have only one photon in the apparatus at a time. We place a detector in one of the beams. We do this many times, changing at random which beam we put the detector in. Since we are blocking one beam with the detector, there can be no interference. The usual quantum explanation is that by peeking we have made a measurement, and come under the domain of the Heisenberg uncertainty principle; however, the peeking occurs before the photon, if it goes through the open slit, strikes the screen; the Heisenberg principle applies to simultaneous measurements, not sequential measurements. Sometimes the detector will register received energy, and sometimes not, the photon in that case presumably having gone through the open slit. In either case, we can say which slit the photon "chose," and thus track the path of the photon.

This means that we have established the moving photon as a frame of reference. We can control when it enters the apparatus, identify its trajectory (which beam it chooses), and determine at what time it is detected, so we can measure its velocity. Will this velocity vary with the motion of the apparatus? (This is forbidden by relativity.) If we produce a Doppler shift, will the amount of the shift be the same for moving source as for moving detector? (This is required by relativity.) The experiments of Relativity Road return to haunt us. The obvious escape from these questions is to repudiate the existence of individual photons, on the grounds that relativity theory prohibits using a light beam, even when quantized, as a measurable frame of reference.

Another difficulty with single photons arises in the delayed decision version of the experiment. Suppose the apparatus is very large. If we do not block either beam, we get interference, so we are dealing with waves. But if we block one beam with a piece of photographic film, the grain of emulsion receives a photon, we say. Now if we decide to block one beam *after* the photon has chosen which beam it is going in, how shall we explain whether it is a particle or a wave? When the wave train divided into twins at the beamsplitter, both twins were wavelike; afterward we found that one of the twin beams was corpuscular. How does the superposition of a wave onto a grain of emulsion change the wave into a photon (corpuscle)? The easy way out is to deny photons, that is, repudiate the corpuscular model for light. If we split the wave train many times, that is, use multiple slits (a diffraction grating) or multiple reflections (a Fabry-Perot interferometer), the unsplittable photon becomes even less credible, compared to the straightforward explanations of classical optics.

Suppose now that instead of dealing with single photons, we consider using *two* photons for each run. They do not have to go through the apparatus simultaneously, but we take a reading after each pair. Again we place a detector in one beam, for example, behind one slit in the Young's apparatus, and again we randomly vary which slit we use. It might be argued that quantum statistics will predict: one-third of the time the detector will catch both photons; one-third of the time it will catch neither photon (null result); one-third of the time it will catch one photon. By classical statistics, we would expect the frequency of these three results to be $\frac{1}{4}$, $\frac{1}{4}$, and $\frac{1}{2}$.

So far, we have considered only the use of light. Suppose now, we consider using *electrons*, one or two at a time in Young's double slit arrangement. (Interferometer mirrors will not work with electrons.) Electrons *do* exist as individuals, and they do not ordinarily interfere. Since electrons (and even better, larger particles) have rest mass, changeable velocity relative to each other and to the observer, and locatability, they do not fall under the relativistic proscriptions that light does. We do not observe interference using electrons, so the paradox of split/no split does not arise. (Electron beam diffraction, discussed on Quantum Road, is a different phenomenon, using a single slit, not two slits.) Electrons do not split, and do not interfere. Since electrons are fermions, and photons are bosons, we might get different statistical results using electrons for the three cases: both caught, none caught, one caught.

If the pair of electrons, or photons, are initially *matched* in some way, such as having opposite spins, this experiment could supplement those used for testing Bell's theorem. One particle could go straight into the apparatus; the other (after initially heading in the opposite direction) could be reflected in after it. We are not here concerned with simultaneous correlations across space, as in Bell's theorem, but with correlations between particles arriving at the same place at different times.

Geodesy and Equipartition in Causal Processes

This section requires somewhat more technical background of the reader than is needed for other sections.

Geodesy, as the term is used in physics, is the tendency of physical changes and processes to take the easiest or minimum path. Almost the whole of physics can be represented in geodetic form. Water running downhill seeks the steepest descent, the quickest way down, and water running into a basin, even one with an irregular shape and bottom, distributes itself so that its surface is as low as possible, the water then has the minimum potential energy in the earth's gravitational field. Light finds the quickest trajectory through an optical system (Fermat's principle of least time). The path of a body in a gravitational field (i.e., free fall in space-time) is a geodesic. Feynman's formulation of quantum mechanics is based on a least-action principle, using path integrals. Maxwell's equations can be derived as conditions of least action. Newton's mechanics is contained in Hamilton's principle of least action, and also in Gauss' principle of least constraint. Thomson's theorem states that electrically charged particles arrange themselves so as to have the least energy. The Second Law of thermodynamics requires that thermal systems change along a sequence of configurations, each having a higher probability of occurrence than the preceding configuration. And there are more.

Geodetic principles are sometimes viewed with suspicion because they seem to imply that the physical process is teleological, as though it has an intrinsic purpose, whereby the sequence of events is determined not by the present state of the system, but by its past and future states. Water, light, and electricity seem to search for shorter or easier routes, as though they wanted to find one, or had advance information that one might be available.

Geodetic behavior can be understood as a causal process, however, without such anthropomorphic or teleological implications. First, however, let us clarify what is meant. As the above example of the Second Law of thermodynamics indicates, sometimes we have a maximum, rather than a minimum. We really mean that when a trajectory (sequence of changes of a system) changes a little bit, some overall property (in the case of Fermat's principle, the total time of travel of the light) does *not* change, at least to the first order. In mathematical terms, we have either a maximum, a minimum, or a point of inflection.

A system does not always find a geodesic, that is, it is not always successful. Sometimes the brook does not find the shortest path down the hill because the geodesic path is separated from the present path by too big a barrier. There are several ways in which a changing physical system, for example, a flowing brook, can ''search'' for an optimum way of reaching the lake, or equilibrium.

First, as explained in the discussion of Fourier components, almost every system can be resolved into waves or harmonics of some kind. Waves have a finite extent in time and space; if one part of a wave front strikes a more favorable (e.g., less resistant or faster) medium than the rest of the wave, the whole wave may deviate correspondingly. Sometimes the wave may be deviated because of diffraction.

Second, there is always noise in physical systems, fluctuations amounting to a sort of nervous twitch. These may lead the system into a different trajectory.

Neither of these factors, however, suggests that the system will ''better itself.'' But suppose that certain directions of the fluctuations or deviations have an advantage, or positive feedback, compared to other directions. Consider what happens when automobiles drive on a road covered with slushy snow; the tires tend to clean the road. This is not due to any desire of the car or the driver, but to the snow being thrown to one side or the other by the moving tires, on an equipartition basis. If the snow is thrown to another portion of the road, it will be thrown again by later cars. But if it is thrown to the edge of the road, or off the road altogether, then it is removed permanently. Thus, eventually all the snow reaches the edge of the road. (See the discussion of random walks, on Probability Road.)

When a brook cuts a shortcut, it tends to keep it, because it picks up greater velocity and current flow more readily than it suffers retardation or slowing down. The key lies in the asymmetrical consequences of the variations in path, shorter paths being "rewarded" more than lengthening paths. In biological evolution, random variations in the DNA molecule (mutations) sometimes have a significant survival value for the species, and are "rewarded." The overall property of the system, such as total action or total time, which is minimized (or maximized) is one which is affected by fluctuations in the system; this overall property tends to settle into a groove, if it finds one, much as a chemical system tends to stop reacting when it reaches equilibrium. To put it another way, some random fluctuations in a system may cause some overall property of the system to stop changing when it reaches a stationary value.

Geodesy may be a general tendency of all physical processes, although it may be masked in some cases by other factors. Consider that a physical system of interest has as many degrees of freedom, or physical variables, as we choose to specify. We can imagine that each of these degrees of freedom constitutes a dimension; thus we specify a multidimensional system space. Each state of the system is represented in the system space by a single point. As the system changes, a trajectory is traced through the system space. A general principle of geodesy can be stated: the length of any actually traveled trajectory in the system space of a physical system tends to be an extremum (i.e., maximum, minimum, or point of inflection).

One difficulty with this principle is that the choice of variables is arbitrary, and some work out better than others. Hamilton's principle requires that we take the difference between the kinetic and potential energies of a mechanical system, as they vary over time. This difference in energies, which is called the Lagrangian, is hardly the first variable one would think to choose.

The trajectory of a process through its system space corresponds to the successive states or changes of the physical system, as it passes from its initial state to its final state. Due to noise, diffraction, and other uncertainties, a trajectory varies and jitters, amounting to a random "searching" process. If the trajectory is quite far removed from a geodesic, the variation in total trajectory length will in general be larger than will be the case if the trajectory is close to a geodesic. Thus, the rate at which a trajectory changes will in general slow down as the geodetic trajectory is approached. Feynman[7]

---

[7]R. Feynman, *Lectures on Physics*. Addison-Wesley. 1963. Vol. II-19 and Vol. I-26. The quantum version of this approach is given in R. Feynman and A. Hibbs, *Quantum Mechanics and Path Integrals*. McGraw Hill. New York, 1965. Especially in Sect. 2-3.

attributes this tendency to a decrease in phase changes as trajectory lengths become more nearly equal (approaching a geodesic), but phase changes, and waves, are not needed for the explanation. The decreasing rate of a chemical reaction, or of a heat flow, as equilibrium is approached is a good illustration. This description can be expressed mathematically as a diffusion equation, which is of the same form as the Schrödinger equation of quantum theory.

The random nature of fluctuations is an example of equipartition. It is a tautology to say that in the absence of an asymmetry, we get symmetrical distribution among possible configurations. In a geodetic process, the asymmetry appears, not in the fluctuations themselves, but in some consequences of the fluctuations. God plays dice with the universe, but he weights them in subtle ways so that they help systems find the easiest (geodesic) way to a final state. As the brook approaches the lake, the gradients fade away, and its trajectory becomes random or serpentine—noisy because dominated by equipartition. Causality is the inference we drawn from *violations* of equipartition; once we discern a pattern in what appears to be chaos, or the random distribution becomes skewed, we start looking for the little green men. In other words, patterns contain information that random distributions do not, and information has physical significance, as we saw in the case of Maxwell's demon.

On the other hand, equipartition is the inference we draw from the *absence* of causality. We say the dice game is honest when there are no causes or patterns introduced beyond the apparently random motions of the dice. Just as deeply as we believe in causality, we believe that in the absence of causality each of the dice configurations will appear with equal frequency. Complete causality we call determinism, and complete equipartition we call indeterminism. A fanciful schematic of the relationship is given in Figure 17.

The holistic point of view advocated here requires a synthesis of the probabilistic (statistical) basis of quantum mechanics and the deterministic implication (Laplace's demon) of classical physics. Integration of the principles of geodesy and of equipartition seems to offer such a synthesis. The tendency of a system to assume its most probable configuration can lead to (a) a tendency to minimize the availability of its energy (the Second Law of thermodynamics), (b) an equipartition principle, and (c) a geodetic trajectory through its system space. Beyond these, however, we need a theory of measurement that will embrace both classical and quantum measurements. This goes beyond the scope of this book. However, an approach is indicated in Appendix C.

The basic difficulty in combining classical and quantum measurements in a single theory is expressed in the complementarity principle. As Fritz London and Edmond Bauer have shown,[8] quantum events are characterized by

---

[8]F. London and E. Bauer, "The Theory of Observation in Quantum Mechanics." Reprinted in *Quantum Theory and Measurement*. Princeton University Press. 1983.

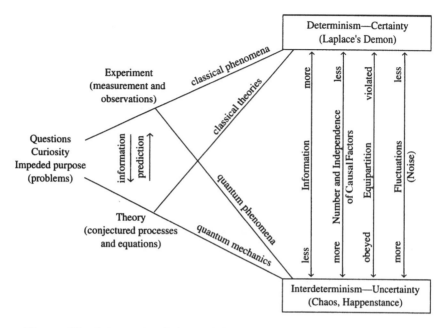

Figure 17 Schematic of determinism and indeterminism in science.

a mixing of the observing (measuring) system with the observed system to such an extent that they cannot be distinguished in the final state. Another way of looking at this situation is that in relativity theory there is a nice symmetry between the observed and observing systems, the two frames of reference in motion with respect to each other; they give equally valid scientific results, and the observer can occupy either, or both, of them. In quantum measurements, on the other hand, the observer must remain in the observing system. The physicist is somewhat in the position of a psychiatrist who must understand, explain, predict, and modify the patient's emotions, without being able to experience them directly.

What the *h* Is It?

The central mystery in quantum theory involves Planck's constant. Its value has been calculated from measurements using different quantum phenomena, as 6.626×10^{-34} joule seconds. The question is: Why do the different methods agree? Why is the "grain size" of our slightly fuzzy universe a universal constant? Why doesn't it vary under different conditions?

Einstein's theory of relativity exchanged absolutes, rejecting Newton's assumptions of absolute rest and absolute time, and introducing a new absolute, that of the velocity of light, c. Quantum theory rejects classical physics' absolute separation between observing and observed systems, and introduces an absolute constant of action, h (Planck's constant). There is a third such absolute constant in the physical universe: the charge on the electron, e (and with reversed sign, the charge on the proton). These three constants, c, h, and e, do not vary with velocity of source or observer in a universe in which almost everything else does, including space, time, mass, force, and energy. The electronic charge does not usually appear alone, because the only way we can detect it is by its interaction with another electric charge. So the fundamental constant is really the product of the two, e^2. The three constants, e^2, c, and h combine to form a dimensionless constant, e^2/hc. This has a value of 0.007297 .... It is a pure number, and its value does not depend on the length of the standard meter, the time of the standard second, or the mass of the standard kilogram.

It is called the fine structure constant, and it arises in quantum mechanical calculations. It is not a trick played with numbers, but a physical constant. Its dimensionlessness has led physicists to conjecture that it may have an undiscovered significance, relating to some general characteristic of the physical universe. The velocity of light may indicate an electromagnetic property of space-time, and h seems to be a fundamental unit of the ratio of electromagnetic energy to the frequency of the electromagnetic radiation, a quantum of action; e may relate to a more fundamental force than the other three forces that physicists hypothesize (gravity, and two nuclear forces). The fine structure constant is the simplest combination of the three velocity-independent constants that is dimensionless, and all three pertain to electricity.

The reciprocal of the fine structure constant is very nearly 137. More accurate measurements show that the reciprocal is not exactly 137, but the numerological itch persisted, and people tried to derive it, or its reciprocal, from first principles on a mathematical and/or metaphysical basis. One such derivation was reported in a French journal; I cannot criticize the derivation, although I read it five times and it is only a few paragraphs, because I did not succeed in understanding it. It reminded me of what happened when the editor of a scientific journal could not make sense of a paper which had been submitted, and passed it on to Wolfgang Pauli for his opinion. Pauli's verdict: ''This paper is not even wrong.''

Some astute people put a computer to work on the problem, and found that there were plenty of other ways of combining integers and mathematical constants algebraically to produce the value wanted, within experimental error. So, we are back to square one with the fine structure constant, and with Planck's constant.

There is more to the mystery of h than the fine structure constant. Planck's

h appeared for the first time in the derivation of his radiation law for black-bodies. It arises as an ad hoc strategem to avoid the ultraviolet catastrophe, and has not had any better birth in all the developments of quantum theory since. Planck's original derivation had objectionable steps; Einstein rederived it, avoiding some of the holes in the road, 16 years later. But h still remains a deus ex machina, whether as a unit of action or of angular momentum.

Quantum mechanics exists in at least four different formalisms: Schrödinger's wave equation, Heisenberg's matrices, Dirac's postulational approach using complex vectors, and Feynman's path integrals (least action). In each of these quite different formulations, h is introduced arbitrarily, empirically. Planck's constant enters into almost every quantum calculation. Sometimes it cancels out, and remains hidden, but it is always lurking nearby. Quantum electrodynamics (QED) centers on incorporating h into formulation of Maxwell's equations. If the value of h is taken as zero, the equations of classical physics appear; that is, quantum phenomena approach classical cases as h is reduced in value. This proposition is called the Correspondence principle of quantum theory.

But the ultimate significance of h, or whether it even has such a significance, remains a mystery. Many physicists think such speculation is a waste of time. h is there; use it, don't ask foolish questions. But some of us insist: predictions without explanations are not enough; there must be more. Yes or No?

10

Road to the Stars

The reader who has stuck it out this far deserves to find easier going on this further road. This chapter requires less technical background. It returns to some of the Big Questions which were asked of the stars in ancient times, that have to do with what force or forces created our universe and whether life on earth is unique or life might exist somewhere else in the universe.

Scientists in Newton's time were interested in much more than the solar system. They were searching, as all scientists are to this day, for answers to Big Questions about the purpose of life and where it comes from. The ancients asked these same questions, though usually in different terms. Our terminology today has become more sophisticated, and we have a lot more scientific knowledge, but the questions are still open. Does God play dice with the universe; are events, including the origin of life, determined by happenstance, accident, and coincidence, or are there assignable causes to everything we observe? Even if we cannot assign causes, is everything we observe in principle intelligible, capable of being understood by us?

This chapter discusses these questions, and uses some of the scientific information we have acquired in modern times, especially some of the things discussed along the Six Roads from Newton. Even so, the answers are never final, never complete. They always raise new questions. The search for where we are and who we are is never ending. I hope it never will end, even though it involves controversy. People who want to avoid controversy shouldn't ask questions. In some laboratories a story is told of the boss (teacher, leader, director) who asks a subordinate to study a problem and report back on it. The subordinate does so, including a strong recommendation for a certain line of action. The boss is annoyed: ''I told you to study it; I didn't tell you to form an opinion!''

Around Home

Let us start with a discussion of how well modern physics explains some of the most ancient questions of astronomy. Astronomy books don't usually show diagrams of the solar system drawn to true scale. If one were to draw the planets' *diameters* on a scale large enough to show how much smaller Mars is than the earth, then, on the same scale, the planets' *distances* from the sun and from each other would make the book enormous. If, on the other hand, they show the planets' distances from the sun in proper proportion on an ordinary page, the planets must be printed as tiny dots.

If the sun were scaled down to the size of an orange, 8 centimeters in diameter, then the earth would be a grain of sand revolving around the orange 8.6 meters away. Jupiter and Saturn would be cherry pits, 45 meters and 82 meters away from the orange. How far, on this same scale, do you think the nearest star to the sun, Alpha Centuri, would be? A mile? 50 miles? 300 miles? The answer is: over 1,400 miles. There's an awful lot of space out there. The sun's actual diameter is 1.4 million kilometers.

The distribution of mass in the solar system is heavily lopsided. The sun, a typical star, is entirely gaseous, yet it has more than 99% of the mass of the solar system. Jupiter has more than twice as much mass as all the other planets and their moons put together. If you take Jupiter out of the accounting, Saturn has more than twice as much mass as all the other planets and their moons put together. That doesn't make the earth too important, except that we have the most water, the most oxygen, and much the nicest weather. We're pretty sure, despite all the science fiction stories, that there isn't any life elsewhere in the solar system, that is, right around home. Elsewhere in the galaxy is a different story, discussed a little later.

All the planets go around the sun in the same direction, and the sun rotates in that direction too. With very few exceptions, the moons of the planets revolve in this same direction, too. And so do the several thousand asteroids, and the numerous comets. The implication is that the whole system was formed at about the same time from a single cloud of gas and dust, with everything swirling in the same general direction. This model also makes it likely that as the sun coalesced from this cloud, residual material formed the planets, and they were constrained to move around the sun in orbits close to the plane of the sun's equator. The orbital planes of all the planets except Pluto and Mercury lie within 4 degrees of the earth's orbital plane (called the ecliptic), and even the plane of Mercury's orbit is only 7 degrees tilted. Pluto may be an adopted child, a late addition to the family, perhaps a comet or asteroid that wasn't too careful.

Why do the planets each stay in one plane as they go around the sun? It would be prettier if they bobbed up and down, like cars on a hilly road. The answer is in Newton's laws, which we discussed in the first chapter. The sun provides a central force field for each planet. Since the vector of gravitational attraction is along a single line, the motion must be confined to a plane containing that line. To go up and down would require a vector with a "vertical" component.

Johannes Kepler inherited careful measurements of the planetary positions made by Tycho Brahe. This was before the telescope was invented. Tycho had tried to modify Ptolemy's geocentric system by having the sun and moon go around the earth, but the planets going around the sun. It was an unworkable compromise. Kepler took up the task, and Kepler, too, was deeply convinced that the Creator had built perfect regularity into the solar system. Copernicus had specified that the planetary orbits were circles, because he considered circles to be perfect, and certainly the simplest, curves. Kepler had to retreat from that simplicity, but he found that elliptical orbits fit Tycho's observations. Kepler's first law states: The orbits of the planets are ellipses, with the sun at one focus. Kepler found that the planets move faster when they are nearest the sun (perihelion) and slower at the other end of the ellipse (aphelion). Kepler's second law states: The line from the sun to each planet sweeps out equal areas in equal times. These laws are illustrated in Figure 18. Since Kepler did not know about gravity, he could not explain the increase in speed as due to the increase in gravitational force, when the planet was nearer the sun. Actually, Kepler's second law expresses the conservation of angular momentum.

Kepler was determined to find the Creator's perfection in the orbits of the six then-known planets: Mercury, Venus, Earth, Mars, Jupiter, and Saturn. Kepler knew, from Euclidean geometry, that there are five, and only five, "perfect solids." The Euclideans had proved, and the interested reader can find the proof in a solid geometry book, that if a solid is bounded only by polygons, and each polygon is "perfect," that is, has equal sides and equal angles, like a square, and each side of the solid is congruent to all the other sides, then there can be five, and only five, such solids. The cube is the best known one, but there are three with equilateral triangles on each side, and one with regular pentagons. That's all there can be.

Kepler had the idea that this elegant situation would have been used by the Creator. The five spaces between the six planets corresponded to the five perfect solids. The trick was to fit the perfect solids in the spaces. Kepler didn't succeed in this, but he did find the connection between the distance of a planet from the sun and its orbital period; the further out a planet is, the longer it takes to get around the sun. Kepler's third law is: The squares of the periods are proportional to the cubes of the average distances.

Newton was able to *derive* Kepler's three laws. (Newton had first to work in the other direction, deriving his laws of motion and the law of gravity from Kepler's laws.) However, Newton did not publish his work.

The story is that Edmund Halley, a young member of the Royal Society, was investigating records of the appearance of the comet that came to be named for him. Halley found the appearances were every 76 years, and guessed that it was the same body. Halley wondered what kind of an orbit it had, but he didn't know how to go about calculating it. So he did the next best thing: he asked Newton, who answered, as the story is told, "an ellipse, of course." Pressed, Newton agreed to "put his notes in order." And so, about 2 years later, in 1687, Newton's masterpiece was published, in Latin, at Halley's expense.

Newton's crisp answer, "an ellipse," came from the following line of reasoning. The law of gravity is a square law. Equations with a square term are conic sections: circles, ellipses, parabolas, or hyperbolas. Parabolas and hyperbolas are not closed curves; a body following such an orbit will go around the sun only one time. To have a circular orbit, the planet has to have been set in motion with perfect x-y symmetry; the chances of this are practically nil. So, an ellipse is what is left. Actually, most of the planets

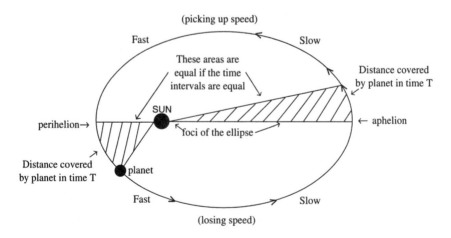

Figure 18 Kepler's first and second laws (From Cutnell and Johnson, *Physics*, 2nd ed., Wiley, New York, 1992, with permission.)

have very nearly circular orbits; comets, however, such as Halley's, have elongated elliptical orbits.

In rederiving Kepler's third law, Newton found that Kepler hadn't got it quite right. The proportionality between the square of the period and the cube of the distance involves the *sum* of the mass of the sun and the mass of the planet; since the different planets have different masses, this changes slightly, although only in the case of Jupiter does it amount to as much as nearly 0.1%. However, when Kepler's third law is applied to double stars, the correction is significant. Here is Newton's result:

$$\frac{(\text{period})^2}{(\text{average distance})^3} = \frac{4\pi^2}{G(M + m)},$$

where M and m are the masses of the two bodies, and G is the constant from Newton's law of gravity.

Newton's work shows that Kepler accomplished more than Kepler had realized. *Any* body revolving around another body, with gravitational attraction between them, follows the same laws. Kepler's second law applies to comets, and ordains that when they pass near the sun, they move very fast, and so are in our view only a few weeks, but they dilly dally for years out beyond Pluto. The moon going around the earth obeys Kepler's three laws (and Newton's, of course), as do the spaceships we put into orbit around earth or around any other planet. Planets, yet undiscovered, revolving around distant stars, must also obey all these laws.

Our solar system is thus not special in the way it was thought to be by the ancients. This determination leads to the natural follow-up question: If our solar system isn't unique, if it follows universal physical laws, how unique are *we*? Why did life begin on earth? How special are we humans? About 98% of the human body is composed of four elements: hydrogen, oxygen, carbon, and nitrogen. The stars, including our sun, are mostly composed of the same elements. If you don't count helium, which is chemically inert, then the sun is composed of about 98% of the same four elements that the human body is. Is this a coincidence? Probably less so than the similarity between our blood and seawater, both of which are salty. The atoms in my hand, and in yours, were probably once part of a star, the one that was around here before our sun was formed. If we are so closely related to the inanimate universe, and if it should turn out that life exists on other celestial bodies, what is our significance?

Where Did We Come From? How Will It End?

The "creation question" is as old as civilization, and probably even older. So are questions about our future; this was the main thrust of astrology.

Scientists have rejected astrology, but not the questions it addresses. Today we answer the scientific portions of these questions in terms of a model of an expanding universe and a Big Bang.

From a scientific standpoint, we start with a calculation of the size of the universe, after which we can estimate how old it is. After that, we can discuss how things will turn out, in particular whether we will freeze when the sun burns out. Finally, we can address the creation question itself: the Big Bang and the Big Crunch.

The modern discussion was launched in the 1920s by Edwin Hubble and others, who measured many galaxies, which are agglomerations of billions of stars, like our own Milky Way Galaxy. Our Galaxy is a disk-shaped swarm about 100,000 light-years in diameter. Other galaxies are of similar sizes, but millions, and even billions, of light-years distant. Hubble measured the Doppler shift from different galaxies, and found that on the average, the more distant the galaxy, the greater its Doppler shift, toward the red. That is, the more distant galaxies seem to be receding from us at greater velocities. Hubble estimated the relative distances of the galaxies by their faintness. All but two or three of the very nearest galaxies, out of the billions of galaxies, are invisible to the naked eye, but Hubble was using the Mt. Wilson 100-inch telescope, at that time the largest telescope in the world.

The excitement, and controversy, aroused by Hubble's findings fueled the campaign to build still larger telescopes, especially the 200-inch on Mt. Palomar. The problem was complicated because the relationship between the distance of a galaxy and its recessional velocity was not constant, but only a statistical finding. That is, *on the average*, it was true—not for each galaxy that was measured. For example, one of the nearest galaxies to us, the Andromeda galaxy, is moving *toward* us at about 300 kilometers per second (a blue shift). (Not to worry; it won't get here for quite a while.)

Hubble's "constant," that is, average, has been revised up and down as we make more measurements. A respectable value is 16 km/s per million light-years. That is, a galaxy that has a red shift (velocity) of 32 km/s can be estimated as being 2 million light-years' distant. A light-year, despite the name, is a distance, not a time; it is the distance that a light signal travels, in space, in one year; this is 5.9 trillion miles, or 9.4637 trillion kilometers.

To estimate the size of the universe, we simply apply Hubble's constant to the largest red shifts we can find. These are associated with quasars, some of which seem to be receding at velocities of at least 15% of the velocity of light. Cranking the numbers together, we get for the distance of the most distant quasars, between 2.5 and 3.0 billion light-years. What's out beyond there? No one knows. Presumably, space and distances don't have the same physical meaning out there as they do around home.

We shall disregard this question, and also the controversies over whether the quasar red shifts are to be interpreted as simple Doppler shifts, and also the controversies over what quasars really are. If we take our calculation at face value, and assume that the universe has been expanding at the same rate since the Big Bang, we can calculate backward to when the Big Bang occurred. If we drive to a city 300 miles away, averaging 60 miles an hour, and arrive at 3 in the afternoon, we must have left home at 10 in the morning. That's all there is to the calculation. We get for a result, using the same Hubble constant, that the Big Bang detonated about 17 billion years ago.

This figure raises an interesting question about the generations of stars. If the lifetime of a typical star, such as our sun, is about 10 billion years, how many generations of stars can there be? The answer is: about 2. Sure enough, Baard found that stars can be divided into two categories, called Population I stars and Population II stars. Our sun is in the younger, or second generation, of stars.

How do stars die? The biggest stars die violently, as novas or supernovas, which are tremendous explosions. But stars the size of the sun become red giants. They expand, cooling somewhat (which is why they turn red). Our sun will expand beyond the orbit of Venus, which means that Mercury and Venus will disappear in the plasma soup. What will happen to us? First the oceans will boil away; then the land will vaporize. We won't be here to freeze when the sun eventually burns out, as it will, because we will have been fried before that. People who believe in hell can take comfort in citing the authority of astronomy for support. For the fearful, this frying (actually, vaporizing) is not scheduled for 5.5 *million* years from now, but 5.5 *billion* years. So let's make the best of it.

Having dated the Big Bang, when the whole universe began to expand from a single speck, let us discuss the event a bit. What ignited it? How did the speck form? What was before then? We can't answer these questions, but a lot of work has gone into proving, or disproving, that there *was* a Big Bang. The evidence consists mainly of two items. First, Hubble's constant, which implies an expanding universe on the basis of observed red (Doppler) shifts. Second, astronomers have measured the residual radiation from the Big Bang. We figure that all that heat didn't completely disappear in the 17 billion years; a very faint amount of radiation, in the far infrared, is still around. The amounts have to jive with the blackbody radiation curve (see Figure 15) for that temperature, which is about 3 degrees above absolute zero. Stated in terms of distance, astronomers figure that the initial heat of the Big Bang is spread out over the universe, with its 2.7 billion light-year radius.

Perhaps the reader shares my scepticism about calculations of what happened in the first second, or billionth of a second, after the Big Bang. How can one place confidence in the equations when one cannot meaningfully ask what things were like an hour *before* that time? The evidence for the Big Bang seems quite thin to support such a weighty conclusion. Perhaps the galactic red shifts have another explanation. Scientific caution suggests keeping an open mind about any model of creationism. Must there always be a beginning?

Astronomers are also interested in what things will be like *later*. Will the universe keep on expanding, or will it slow up, perhaps reverse itself and contract, becoming (ending in?) a Big Crunch? The discussion hinges on the amount of dark matter "out there." This unseen matter may be in the form of dead stars, or black holes, or subatomic particles; nobody knows but it's fun to guess.

Why do we assume that there was only a single Big Bang? One reason is that all the matter we observe seems to belong to the same type. Every type of particle has its antiparticle; that is, matter can exist in either of two symmetrical forms. Why don't we find them both in approximately equal proportions? Atoms and compounds can be built with antimatter, just as well as with matter. However, when a particle meets its antiparticle twin, when we make them meet in the laboratory, they annihilate each other. The two masses disappear, creating electromagnetic energy in accordance with the relativity equation, $E = mc^2$.

The implication is that atomic matter was formed only once, or at most a few times. Matter won the toss over antimatter, supposedly in the first second or so after the Big Bang, and that settled it. If matter and antimatter had been created from electromagnetic energy (the annihilation-creation process is reversible) many times, half of the time antimatter would have won out (equipartition). The great prevalence of one form over the other today is evidence for a single Big Bang. When antimatter is created today, it is soon annihilated.

Biology may have a parallel to the matter/antimatter symmetry in physics. It seems that life on earth is characterized by DNA molecules, which have a left-hand thread. A right-handed helix would work just as well, but as with contact between matter and antimatter, contact between DNA molecules of opposite parity would be disastrous for both. If each type of DNA were equally likely (equipartition), then life probably formed on earth only once, or at most, a few times.

Does Life Have Purpose if Inanimate Processes Do Not?

If life is something that is likely to happen when conditions are favorable, does that mean that life has no more meaning than volcanoes, tides, and other natural phenomena? Scientists are prejudiced against exceptions; we prefer to bet with, not against, the odds. If life does not exist anywhere else in our Galaxy (which has at least a hundred billion stars), if we are the Great

Exception, the Only Child, if the odds against life originating even under favorable conditions are tremendous, then a scientific explanation for the origin of life is more difficult than if we assume the opposite. In either case, we have not succeeded in explaining it yet.

As scientists, we can argue the processes of biological evolution, and the role that accidental mutations may play. We can also ask: What is the probability of life developing from inanimate materials? Not only the first cellular life, but the whole evolutionary development of successively more complicated species, seems to demand some explanations, perhaps some principles, which have not yet been discovered, or if claimed, have not yet been widely accepted. Even if we modestly downplay the emergence of homo sapiens as the apex of evolution, the number of factors that must fit together to make life and evolution of species possible, is awesome.

In order for birds to evolve from reptiles, they not only had to develop feathers, but also lighter bones, different muscles, a shift in center of gravity, control of body temperature, and metabolism changes. These changes had to be more or less coordinated; it is very unlikely that they developed one after the other, independently. What principles guided the changes? Some people see what they want to see in such puzzles, for example, God. Others look for survival advantages in the results of random mutations, somewhat as passing cars clean the snow off the road, as discussed in the section on geodesy in Chapter 9. Massive bodies have inertia of motion, temperature, and other properties. Perhaps the tendencies of living organisms to perpetuate themselves, individually and as a species, represents a type of inertia. Life may be created artificially in the laboratory in the not too distant future; scientists believe, or hope, that what we create ourselves, we can better understand.

As for life existing elsewhere, most astronomers are inclined to believe that there are at least several dozen places in our Galaxy where some form of life exists. (Note: this is not equivalent to believing in visitations from such life.) One way of organizing the discussion, and calculating the number of stars that may have inhabited planets orbiting them, is to use the equation devised by Frank Drake, a Cornell astronomy professor. This takes the number of stars that are not binaries (because planets tend to get captured if there are two stars close together), and multiplies this number by the average number of planets. This number is then multiplied by the fraction of planets which have liveable climates and atmospheres, and then by the fraction on which life has actually developed, and then by the fraction where life is intelligent (by our standards). Then we have to consider how long an average civilization lasts. Earth may not be the only place where civilization, and perhaps all life on a planet, can terminate; there may have been life in places where it no longer exists. Each of these numbers has been debated, with the final number ranging between 20 and a million.[1] The reader can decide for himself whether the higher estimates of extraterrestrial life were made by the optimistic astronomers or by the pessimistic astronomers. But either way, most astronomers reject exceptionalism.

Pioneer 10, launched March 3, 1972, was a lighthearted way to search for life elsewhere. It was the first spacecraft to be sent outside the solar system. The implicit message, I suppose, would be: Having wonderful time; wish you were here, but it couldn't be written in words or letters. The big problem was the return address. For a reproduction of the plaque, and explanation, see Carl Sagan's *The Cosmic Connection*. There was some excitement concerning the "signatures," a nude man and woman. Some earth people objected that the first extraterrestrial message was pornography!

Drake's equation starts us toward answering the question of whether life is "ordinary," "common," "to be expected." It gives little help as to the ratio of causality to accident in the origin of life, or in the evolution of higher forms of life from simpler forms. Perhaps physics will get some help from biology in dealing with the ratio of causality to accident, which we wrestled with in Chapter 9.

Does God Play Dice with *Us*?

Life has a purpose for us humans because purpose is a human trait. Virtues, crimes, good and bad morals, are *ours*. The central importance of the experimenter in the relativity and quantum theories appears also, in different form, in ethical theory. The moral significance of behavior depends on both our intent and our actions, somewhat as a measurement depends on both the signal and the detection.

Ours is a time of floating faith; like straw on water, it will attach itself to almost anything with which it comes into intimate contact. A bumper sticker expresses the quiet desperation: "God said it; I believe it; that settles it." Science teaches that a questioning attitude is a virtue, and that thinking is one of the roads to morality. But science hasn't found "the answers" to the Big Questions either.

A scientific absolutist is a determinist, believing in predestination and Laplace's demon. In an objective morality everything is either forbidden or required. If we do not have free will, if we have no choice in our behavior, how can anyone be praised or blamed? The determinism of the Newtonian Great Clockwork is amoral; there is no moral responsibility. If criminal

---

[1]G. O. Abell, *Exploration of the Universe*, 4th ed. CBS College Publishing. 1982. Chap. 36: "Life in the Universe: Are We Alone?"

behavior is attributed to (caused by) childhood abuse, or poverty, or inherited genes, how can we blame, or punish, criminals?

But the *in*determinism suggested by the conventional quantum theory is also amoral. If we do have free will, what is it free *from*? Everything? The case for free will cannot rely on Heisenberg's uncertainty principle because (1) it applies only to certain pairs of measurements, not to all, and (2) uncertainty is not a license for either human or divine purpose, or will, but only for ignorance, a restriction on possible knowledge.

A first step out of the determinist/indeterminist swamp was mentioned above; it is to recognize that even in a causal chain of events, *we* are one or more of the links in the chain. Whether we have free will or not, we can be, and are, causative agents, gears in the Great Clockwork. Just as the scientist is part of the measurements he makes, so the person is part of the effects produced by his own behavior. Archimedes said that he could move the earth, if he could find a place off the earth on which to put his lever; he couldn't, and man the moral creature has a similar disability. It does not make sense, scientifically or morally, to presume a point of view completely independent of, and detached from, everything that happens. People are free to search for morality and knowledge, including looking into the mirror.

A second step out of the intellectual swamp is to recognize the significance of multiple causation. There are many contributing causes to our behavior, some in series, some in parallel, some loosely connected, some essential. Complicated networks sometimes take on a meaning of their own, for example, synergism. Causes can be fuzzy, multifaceted. In practice, the whole is usually different from the sum of its parts. Just as individual molecules striking a wall in huge numbers become uniform pressure, so a large number of interrelated factors may acquire meaning from their patterns. Emotions as well as clouds form patterns.

A third step out of the intellectual swamp consists of recognition of Mr. In Between, of halfway houses lying between determinism and "pure" free will. Instead of seeing every event as either Caused or Uncaused, as Aristotelian logic would have it, we can see events as slightly out of focus, a little fuzzy. Everything that happens is caused, but there are accidental happenings, degrees of potentiality and possibility.

Recognition of unpredictability need not open the door to the supernatural, but merely to uncertainty. The limits on causality are mainly due, not to spooks, but to noise and to the limited accuracy of our scientific concepts and definitions, and to our mistaken presumption that the physical universe is as tight as our mathematical descriptions of it. Laplace's demon was based on this mistaken presumption.

Kurt Gödel, a mathematician of our own century, proved that a mathematically tight set of axioms and postulates *must* contain a logical glitch. If not an outright inconsistency, then the system must at least contain an indeterminate case which cannot be handled. The standard example of Gödel's theorem is: In a town with only one barber, the barber shaves all those, and only those, who do not shave themselves. So, does the barber shave himself? If he does, he doesn't. And if he doesn't, he does.

Where Are the Roads Leading?

In scientific research many, probably most, roads are dead ends. They are generally forgotten. Some few are worse, leading into swamps of untruth. Some racial theories have done that; astrology and palmistry seem to fit that description also. A few, such as the wonderful roads blazed by Newton, Maxwell, and Einstein, have led us to new worlds, more remarkable than the new world which Columbus led Europeans to. This book has taken the reader on quick trips along wonderful roads of great discoveries in physics; there is a lot more to see.

We are still building roads, of course. That is a characteristic of modern times, even though no one can be really sure just where the roads will lead. Science is part of our zeitgeist, one of the ways in which civilization expresses itself. The building of such roads, that is, the directions in which we ask questions and pursue research, is part of our belief in Progress, the goal of collectively improving ourselves and our lives. Progress has replaced the emphasis which in earlier historical periods was placed on tradition, on continuity and transmission of inherited values and customs. In the course of building a new civilization, we have created ourselves.

Those who followed in the footsteps of Newton focused on forces and masses interacting according to immutable and perfectly exact laws. Those who followed in the footsteps of Maxwell focused on fields and the signals and information that could propagate in them. Those who followed in the footsteps of Einstein focused on space-time. Quantum theory was still more radical, focusing on the strange rules governing atomic events.

Where are we focusing today? In our own time, we have "harnessed the energy of the atom." We are talking of permanent stations on the moon and elsewhere in space, and of information highways that will keep us in constant communication with everything and everyone. We are altering and cloning genes, and it looks as if we will be able to manufacture, or custom tailor, living organisms. Are we prepared to wield these new god-like powers wisely? The New World that Columbus opened up for Europeans did not bring out the best in European civilization. Einstein himself regretted the role that his relativity theory played in leading to the atomic bomb; he said that if he had known, he would have become a plumber instead.

The new powers and opportunities for progress that science places in our hands carry responsibilities and potentialities for harm. Progress always has its price. It is not so much a new malevolence that we have to fear, as of opening up new opportunities for the malevolence we have always had. The automobile, the computer, and modern medicine represent progress, but they also impose on our society a responsibility for traffic congestion, drunken drivers, fraudulent computer practices, and medical malpractice. In the past, many scientists have sought to evade responsibility by erecting a facade of impartiality. There is the story told about Dr. Vilhjalmur Stefansson, the Arctic explorer, who wrote that the Eskimos' diet was almost exclusively meat. Some dieticians challenged this, claiming that such a diet would not sustain health. A controversy grew, so the next trip that Stefansson made into the Arctic was preceded by thorough physical examinations for himself and his colleagues. They then lived exclusively on meat for an entire year. On their return, they were again examined, and found to be in good health. Afterward, Stefansson was telling of the Eskimo diet in one of his lectures, adding that he himself had subsisted on an all-meat diet. Someone in the audience asked, "But Dr. Stefansson, what were you trying to prove?" Forgetting the controversy that had led to the experiment and given it real meaning, Stefanson pontificated, "A scientist does not try to prove anything; he merely seeks to learn the facts."

If scientists close their eyes to the implications of what they consider "pure" research, that is, pretend that it is pure of applications and consequences, applications and consequences will come anyway, and are more likely to be unwelcome. Those who search for new worlds and new truths need to take care that proper use is made of them.

A 13-year-old girl had been so deprived of human contact since infancy that she could not speak. Her condition excited linguistic and medical scientists who saw a wonderful opportunity to test some theories of language acquisition and brain development. They did, to their own advantage and the girl's detriment. The girl had begun to make some progress under individual care; they took her away, tested her for a time, and then when the funding ran low, cast her off, even suing her for their "treatment". Scientists have no procedures either for preventing such exploitation or for taking action when it has occurred. Scientists have yet to decide what is within and what is outside their responsibilities.

When we pull our car over to the side of the road upon hearing the siren of an emergency vehicle, we think, "It is the law." Then we think, "Maybe next time, I or someone in my family, will be needing emergency assistance." And then we realize, "It is good to live in a community where people give priority to those most in need." The first two reasons are logical; the third goes further. One of the scientists of Newton's era, Blaise Pascal, wrote, "The heart has reasons that the head knows not of."

Heinrich Himmler, the Nazi chief of the SS, once wrote that nothing should be done for its own sake. Here the devil gives us a clue. We should pursue more of our activities, including the sciences, the arts, and sports, for their own sakes, wholeheartedly, not for what we hope to get out of them. The best activities are both ends and means; too often, instead of being a labor of love, what we do has an ulterior motive. No wonder nice guys so often finish last in our society. In science, there is a tension between those demonstrating their intellectual prowess and aiming at prestige, and those who are driven to find what keeps the ducks' feet from freezing, that is, by curiosity.

Our society does not make the best use of the brains and talents at its disposal. Effort is made to turn every human talent to profitable account. The artist is turned into a commercial artist; the musician plays routine music year after year; the writer becomes a copy man; the scientist, a weaponeer; the persuasive person, a salesman; the intellectual, a staff man; the playwright and actors, entertainers. Nothing goes to waste, and everything goes to waste.

As civilization goes, progressive or reactionary, so goes science. Scientists share the malaise of our time; physicists show it in their readiness to surrender faith in the intelligibility of the physical universe (Einstein called it "intellectual resignation") and in their willingness to subordinate their work to the military establishment. We have not solved the riddles of quantum theory yet, but we should have faith. Like Columbus, modern scientists have arrived in a strange new land, which is not well understood. "We are a long way from home" is the way J. Robert Oppenheimer, top scientist of the atomic bomb project, put it. Newton said that he saw further than others because he "stood on the shoulders of giants." We have many more giants on whose shoulders we stand. We have seen, on the six roads from Newton, some of the great discoveries in physics which these later giants have made, and there is a long way ahead. The scientific revolution is not over.

Appendix A
What Is Energy?

Energy is like the Old Man of the Sea in the Sinbad stories, changing its form bewilderingly. The headings of the rows and columns in Chart A1 list the major energy forms. If we start with a form of energy listed across the top of the chart, it can be transformed into another form of energy, listed down the left-hand margin of the chart. For example, electrical energy can be transformed into radiant energy by a light bulb, or a cathode ray tube, or an x-ray tube, or by lightning. The chart thus indicates how energy inputs can be transduced into energy outputs.

Mechanical energy occurs in two main forms, that of motion (kinetic), and that of position in a gravitational field (potential). A pendulum, as it swings back and forth, alternately transforms potential energy (at the top of its arc) to kinetic energy (when it has maximum velocity at its lowest position), and back again. Energy is conserved, but as the pendulum loses its "pep," the energy becomes unavailable, being converted into heat by friction. This is an example of the Second Law of thermodynamics, discussed on Field Road.

All forms of energy ultimately degenerate into heat, which is why all the squares in the bottom row are filled. Some of the unfilled squares in the chart can perhaps be filled in by the reader; some of them, however, may

Chart A1 Energy Transformation Chart

From → To ↓	Electrical	Radiant	Chemical	Kinetic	Potential	Sound	Magnetic	Nuclear	Heat
Electrical	Transformer	Photocell Thermocouple	Battery	Generator Tidal power Quartz crystal	Hydroelectric dam	Microphone	Magneto	Beta emission	Cathode Thermocouple
Light or Radiant	Light bulb CRT Lightning X-ray tube	Fluorescence Laser	Firefly	Meteor			Zeeman effect Faraday effect	Atomic bomb	Incandescence fire
Chemical	Charging a battery	Photosynthesis Formation of fossil fuel		Kneading dough					Endothermic reactions
Kinetic	Motor Galvani's frog	Radiometer	Rifle Internal combustion engine Muscles	Flywheel	Air rifle Pendulum Bow and arrow Sledge hammer	Ultrasonic cavitation	Magnetostriction	Atomic bomb	Steam engine Popcorn Wind
Potential	Capacitor Forktruck	Excitation of atoms	Making explosives	Pendulum Winding a spring			Induction		Hot air balloon
Sound	Loudspeaker Thunder		Firecracker	Drum	Alarm clock	Hearing aid	Audio cassette player	Atomic bomb	Boiling
Magnetic	Electromagnet Solenoid			Cyclotron Synchrotron		Audiocassette recorder			
Nuclear	Cyclotron	Particle creation	Oxidation Combustion Digestion Heat of solution	Meteor impact	Meteor	Focused sound	Hysterisis	Atomic bomb	Nuclear fusion
Heat	Toaster	Burning glass Solar collector						Atomic bomb Fusion	Melting/freezing

represent inventions not yet made. Some of the entries are really multiple transformations; for example, a steam engine transforms the chemical energy of coal into heat in the form of steam, and then uses the steam to produce kinetic energy.

Some physics books define energy as "the capacity to do work." Usually work is illustrated by manual work, such as lifting weight in the earth's gravitational field, for example, shoveling dirt up into a truck. The definition is misleading, because a given amount of energy in the form of heat cannot do as much work as the same amount of energy in the form of electricity. If you run 20 miles around a level track, you do very little work, but dissipate a lot of heat. Your body's energy may be exhausted.

A documentary film features three eminent scientists discussing "What is energy?" George Wald and Linus Pauling are Nobel Prize winners, and Philip Morrison was one of the stellar physicists on the Manhattan Project. These three experts make many interesting comments about energy, but as you listen, you realize that they do not know what energy is. Like them, Chart A1 only skirts around the question.

Appendix B
Some Impossibilities Discovered in Our Time

One of the advantages of scientific methods is that they enable us to learn from our failures as well as from our successes. In the shadow of every scientific achievement, and of every scientific mistake, our knowledge of what can, and what cannot, be done grows. Processes formerly seen as entirely independent are recognized as related to each other, perhaps as special cases under the same theory. Some goals that seem impossible of attainment become possible, and other goals, which we hoped to reach, are seen as impossible. So, while science grows in power, it also finds more and more feats that are unachievable. In short, as we succeed, we learn that certain things won't work.

Here are a few of the sights pointed out along the six roads from Newton, reformulated into negative terms—"you can't do that" terminology.

It is impossible for a device to produce more work or energy than is put into it. Conservation of energy means that perpetual motion is impossible.

It is impossible to convert heat energy into kinetic or potential energy with an efficiency greater than the ratio of the temperature change to the higher temperature (Second Law of thermodynamics).

It is impossible to detect absolute motion between two inertial frames of reference (Newtonian principle of relativity).

It is impossible to use an electromagnetic beam as a frame of reference; therefore, the velocity of light is the same for all observers. Note that this invariance does not apply to the frequency or wavelength of the light.

It is impossible for a body of finite rest mass (inertia) to reach or exceed the velocity of light (theory of relativity).

It is impossible to detect any difference between the inertial mass of a body and its gravitational mass (principle of equivalence, general relativity).

It is impossible to propagate a beam of electromagnetic waves that does not carry energy and momentum ($E = mc^2$).

It is impossible to eliminate noise (fluctuations of magnitude) from any physical system. Thus, it is impossible to exactly repeat, or exactly reverse, any process.

It is impossible to achieve a perfect vacuum, absolute zero temperature, frictionless motion, perfect randomness, perfect periodicity, perfect monochromaticity, etc. Perfection is possible in mathematical systems, but not in physical systems.

It is impossible to measure momentum and position of a particle simultaneously with combined (product) accuracy less than Planck's constant (Heisenberg uncertainty principle).

It is impossible for two electrons in the same atom to have the same quantum numbers, that is, physical state (Pauli exclusion principle).

It is impossible to specify a physical system completely without specifying how it can be measured at least in principle. (This is controversial.)

It is impossible for a scientific statement to be exactly true or completely false (controversial).

It is impossible for a physical system, for example, an atom, to exist in each of two energy states which are closer together than one quantum.

It is impossible for two physical systems in the same initial state to have different outcomes without some difference in the conditions or process. This causality heresy amounts to a question of how accurately we can determine whether systems *are* in the same state.

It is impossible for a physical measurement to occur in zero time or in infinite time. Therefore, perfect resolution and complete information are unattainable.

Appendix C
Measurement as a Creation of Information

For our purposes, a measurement process can be divided into a preparation phase, a detection phase, and an interpretive phase. Each phase can be viewed as the interaction, or superposition, of two systems. The preparation phase involves interactions between the K (Knowing, or mental, system, that is, the observer himself) and the D (observeD) system to be measured. First, there is recognition of a question or problem, and a need or desire for an answer or for more information. Then there is a choice of method, a game plan, a match of question to hypothesis, so as to distinguish between alternative possible outcomes. The preparation phase ends as the plans jell, become more specific and complete as to how, which equipment, when, where, and so on.

The detection phase begins with the alignment of the D system and the G (observinG or measurinG) system in time and space; wires are connected, lenses focused, impedances matched, amplification provided (if needed), instruments calibrated, and so on. Then the instruments are turned on and there is an exchange of energy and entropy between the D and G systems, usually approaching a steady state. This partially merges the systems, sometimes irreversibly, so that their separate contributions cannot be distinguished.

171

The heart of the measurement process usually involves a transduction (see Energy Transformation Chart A1), so that two quantities of the same degree of freedom (variable of the systems) can be compared. The detection phase ends with these two quantities superimposed, so that the reading can be taken.

The interpretative phase involves interactions between the merged systems, and the K system. When the reading is taken, a conscious act, the information is processed; a film may be developed, a reading recorded, data points plotted, and so on. The phase ends with a judgment, for example, "satisfactory" or "snafu," and a decision whether the measurement process should be terminated, repeated, or changed.

Consider a simple measurement of temperature using a mercury thermometer. Having obtained a suitable thermometer, we immerse the bulb in the medium. The thermometer is a transducer, the change of volume of the mercury changes the length of liquid column in the glass tube. When the rate of change becomes suitably slow (steady state), signifying that the mercury has become thermally part of the medium being measured, we note the position of the column relative to the graduations on the glass tube. Thus two lengths (of mercury and glass) are compared, superimposed. Lastly, we decide whether the procedure has been satisfactory, and whether enough information has been created.

The preparatory phase might be a mother concerned whether her child is sick. In the detection phase, she sticks the thermometer into one end of the child, and awaits a steady state. After reading the thermometer, she makes a decision. Note that the most irreversible part of the process may be in the K system—in the mother's mind.

There are, broadly, three kinds of measurement: active, passive, and indirect. In active measurement, energy and entropy are exchanged between the D and G systems. For example, we take a photograph, capturing some of the light that has been bounced off the D system. In passive measurements, energy and entropy are exchanged between some offshoot of the D system and the G system. For example, measurements on starlight do not affect the star, but they do affect the light. An indirect measurement is made in two or more steps. We match up two parts of the D system, separate out one of them, and make our measurement on that one part. This is what the EPR paradox, and Bell's theorem, deal with.

Quantity of Information Created

Information about the D system (which is the one we are interested in) is strained through the G system, that is, it is convoluted, in the sense of the convolution theorem of Fourier optics. It is obvious, especially when we forget something, or lose some data, that information is not conserved. Less obviously, the quantity of information represented by some data is partly subjective and/or depends on the context.

For example, the probability of throwing exactly one 7 in three throws of the dice is 75/216. If I throw a 7 on the first throw, the odds shift to 150/216, so that the first throw has an information value of 150/216 minus 75/216, or 75/216, which is the change it produces in the betting odds. Suppose, however, that the bet is whether or not I throw a 7 in *two* throws (instead of three throws); now the information value of my 7 on the first throw is 30/36 minus 10/36, or 120/216. The same throw of the dice (a 7 on the first throw) has a different quantity of information in the two cases.

The variable information value of data is usually more complex in physical measurement than in dice games. As we see in detective stories, a clue contains more information for the Great Detective than for the ordinary mortals in the story. In general, the quantity of information depends in part on the way the information ''fits'' contextually, on a sort of coherence. The concept of coherence in optics is analogous; both relate to a superposition process (one of waves, the other of data). In the dice example, above, the superposition of the data (the throw of a 7) was with a previous bet, i.e., other data. But in real life, including laboratory life, one of the systems being superimposed may be the K (knowing) system, e.g. the mind of the Great Detective. Thinking, too, is a method of creating information.

In the dice example, we assumed that

$$I = P_2 - P_1 = \Delta P, \tag{A1}$$

where I is information and P is probability. Probability is dimensionless, and always lies between zero (impossibility) and unity (certainty). Information can be expressed in different units; in computer work we use bits and bytes. A picture on a TV screen, or a photograph, uses picture elements, or pixels, the unit being determined by the smallest grain size (resolution) and the number of separable shades of gray (intensity scale) in each. Thus, it makes sense to say, ''This picture is worth 9,234 words.'' If we wish to express information in dimensionless units, in order to express it in terms of its effect on the probability, we can write:

$$\frac{I}{I_{max}} = \Delta P. \tag{A2}$$

Here I_{max} is the maximum amount of data, beyond which further data points, or higher resolution, would not affect the probability significantly.

Generalized Bandwidth

In exorcising Maxwell's demon, Szilard and Brillouin[1] identified information with negative entropy, implying that the unit of information should be joules per degree temperature, like Boltzmann's constant. The information content of physical measurements usually has the dimensionality of the measurements themselves. Here we can use Fourier methods to attack the general problem.

If the voltage (or current) in a circuit is a measured function of the time, $f(t)$, the information can be transformed to the frequency domain (from the time domain) by expressing $f(t)$ as a Fourier series, and noting the coefficients of the different terms: the relative amplitudes of the different harmonics, or Fourier components. Usually, the amplitudes of the components decrease as their frequency increases. When the amplitudes approach, or fall below, the noise level, we judge that we have reached the bandwidth limit of the signal, and perhaps also of the electrical system. If the voltage versus time curve has sharp corners, or spikes, the highest frequency components will contribute significantly there (see Figure 13).

The same Fourier methods are applied to optical images, where a densitometer or intensity scan of the image (through a slit or pinhole, or using a well-focused spot) gives the density or intensity as a function of the distance, x, along the scan line. In this case, the data curve is a function of x, rather than t. Again, the Fourier components of the experimental curve give valuable information about the image quality. The Fourier components are called spatial frequencies, and are measured in line pairs per millimeter (instead of per second). A line pair is one black line and one white line in a test target, and we call the pair, a cycle.

This Fourier method is not restricted to time and space, but can be applied to any degrees of freedom of a physical system, yielding the "bandwidth" in cycles per the degree of freedom we are interested in. For example, we can apply the method to pressure versus temperature charts, or to field strength plotted against voltage. There are mathematical conditions of continuity required, but these are very liberal. The method can even be applied to more than two degrees of freedom at once.

If our interest centers on the performance characteristics of our measuring instruments, or on our experimental procedures, the modulation transfer function (MTF) can be used to indicate the falloff of amplitude with frequency. We repeat, "frequency" can mean cycles per any degree of freedom we choose.

The Whittaker-Shannon sampling theorem[2] shows that the number of

[1]"Maxwell's Demon," see Footnote 1, Chapter 4, above.
[2]J. W. Goodman, *Introduction to Fourier Optics*. McGraw Hill. 1968. Sect. 2-3.

"samples" (amount of information) required to preserve all the information in our experimental curve is given by twice the bandwidth. Note, this is not a pure number of data points, but samples per unit degree of freedom. What we are saying is that our plotted data curve has a finite width (probable error). We do not need more than a certain number of "fat" data points squeezed alongside each other in our "fat" line, because such points would be redundant. By establishing the bandwidth of our data curve, we find at once, from the sampling theorem, what its information content is. If three degrees of freedom are involved, we have samples per unit area.

Determination of quantity of information thus has two steps: (1) Determining the bandwidth of the experimental curve, so that the sampling theorem can be applied, and (2) converting the indicated number of "samples" to nondimensionality so that we can relate them to the changes of probability, as indicated in Equation A2. Expressing information in terms of probability change enables us to relate classical measurements to quantum measurements, because quantum measurements are expressed in terms of the psi function (of the Schrödinger equation) which when squared directly corresponds to probability.

Quantum Measurement

Classical physics focuses on physical processes, while quantum physics focuses on physical states. The limits of classical physics appear in cases where the physical process of interest is significantly disturbed by the measurement or detection process. The limits of quantum physics appear in cases where we inquire about the physical process, demanding more than the probabilities of different outcomes. Both classical and quantum viewpoints regard measurement as a superposition of the observeD (D) and observinG (G) systems, but one considers the interactions of D and G as secondary, and the other considers the interactions as introducing indeterminancy.

There is no generally accepted theory of measurement in quantum mechanics.[3] But two features are frequently included:

1. The process of measuring transforms the D system into a definite state, which it had only potentially previously. This is the usual comment on Schrödinger's cat.
2. The indeterminancy introduced by the measurement is due to a loss of phase information. Lost information is the root cause of irreversibility of physical systems.

---

[3]M. Jammer, *The Philosophy of Quantum Mechanics*. John Wiley & Sons. New York, 1974. Chap. 11.

The first feature has led to much controversy, including that of Many Worlds, and the paper of Einstein, Podolsky, and Rosen (EPR). The second feature is summarized by London and Bauer.[4] "Causality is no longer applicable, it is true; but the reason for this fact is not the impossibility, in the last analysis, of reproducing identically the conditions of an experiment. The heart of the matter is the difficulty of separating the object and the observer."

A classical case involving recording versus loss of phase information is that of holography (see Wave Road). When we take an ordinary photograph of an object, we record on the film the intensities of the waves coming from each point in the object, but we lose the phase relations between these waves. In a hologram, this phase information is recorded, with the result that we can see much more in a hologram than in an ordinary photograph. We also need more fine-grained film for the hologram, because much more information is being recorded in the same area.

---

[4]F. London and E. Bauer, "The Theory of Observation in Quantum Mechanics." Reprinted in *Quantum Theory and Measurement*. Princeton University Press. 1983. p. 220.

Appendix D
Physicists Shouldn't Take Math Too Seriously

When we were in grade school, we thought that mathematical proofs were the ultimate in establishing Truth and credibility. If only political and moral questions could be settled so definitively! But how can one be absolutely sure, beyond any shadow of doubt or later wisdom, that a proof is correct, has no mistakes or oversights in it? We can't, of course, and the history of mathematics and science is replete with defective proofs.

The most important reason for proving a theorem or deriving an equation is not, usually, to establish the honesty or accuracy of the prover, but to indicate which concepts and facts underlie the conclusion. That is, a proof answers the question: On what does this result depend? That is why alternative proofs are often of interest; they show that particular items may, or may not, be essential to the final conclusion.

Mathematics deals with abstract quantities, not physical quantities. Remember when the teacher had us say how many apples she would have, if she started with five apples and two students each brought her an apple? Sometimes she'd ask, "If I had seven oranges, and I ate three, how many oranges would I have left?" And then one day she gave our innocence a fatal blow. She asked, "How many are five and four?" Five and four WHAT?" we shouted. But the damage was done, never to be undone. We were up the mathematical creek, without apples or oranges. And so it continued with more blows to our innocence. She put numbers on top of numbers, and called them fractions. Then she used letters instead of numbers, and called them unknown numbers. Then she had numbers lower than zero, negative numbers. On the Field Road we saw numbers with two heads, vector numbers. As you advance into mathematics, it gets worse and worse.

Kronecker said that God made the natural numbers but everything else is the work of man. That is, he pleaded guilty to all the extensions of the idea of number, but he believed that the positive integers were "natural," that is, there before man. The reader may reasonably dispute this claim, starting with asking where "there," as in "there before man," is. A more fruitful question, which gives a hint where "there" might be, concerns the difference between mathematical concepts and the related, or parallel, physical concepts. Two things equal to the same thing are equal to each other in mathematics; however, two metals in electrical equilibrium with the same electrolyte may generate a current when connected to each other. Is multiplication a shortcut way of counting the little square tiles in the bathroom floor, using only the counted number of rows and columns to get the total number of tiles in a rectangle, or is it an abstract operation? Multiplication can be defined as an operation in mathematical logic, or as a physical process of successive additions.

Three common types of mathematical numbers show the dichotomy between the mathematical and physical way of approaching quantitive problems: irrational numbers, infinite numbers, and imaginary numbers. Let us look at irrational numbers first. If we make two measurements of the same degree of freedom of a physical system, say its voltage, we can always take the ratio of the two voltages. Mathematically, this may not be so; two quantities may be incommensurable, having no ratio at all. The simplest example is the square root of 2. We can show that the square root of 2 is a number that cannot be expressed as the ratio of any two integers, no matter how large. Or consider the problem of writing the ratio of the diameter of a circle to the circle's circumference, or the hypotenuse of an isosceles right triangle that has legs of unit length. Geometers say that they can imagine cutting up a circular area (for example, a pie) into sectors (radial portions) such that the cutting does not come out even, *no matter how small* they make the angular sectors. To a hungry physicist at the table, awaiting his piece of pie, this is nonsense. In other words, the mathematical concept of incommensurability, or of irrational quantities, does not have any parallel in physics.

Or does it? Consider the trouble we have had with arranging leap years, so that the earth is always at the same point in its elliptical orbit around the sun at the same second on the same calendar day. You have probably read how Pope Gregory XIII decreed that the day after October 4, 1582, was October 15th (causing disputes over rent, interest, and paychecks), and how the year 1900 was not a leap year. Did Pope Gregory's reforms make everything perfect? On earth there is no such thing as perfection, only in heaven and in mathematics, each being the work of man. Recognizing this, today we cheat, that is, the astronomers decide when to give their atomic clocks a twitch, to restore synchronism. Thus, we recognize that the day and the year are incommensurable *in practice.*

Infinite numbers are another glitch between mathematics and physics. Are there more even integers than there are even plus odd integers together? Well, the number of even integers is countably infinite, and so is the number of integers, so they can be placed in one to one correspondence with each other. That is enough to convince mathematicians, apparently, but not me. Mathematicians show that the number of rational numbers is countably infinite, but the number of irrational numbers is not, and that the number of points in a line segment is not countably infinite. Every department on the campus has fun in its own way, as it should.

Now consider the physical view of infinity. When we say that the image formed by a lens is at infinity, we mean that an image is not really formed, but can be considered as formed so far away that the other dimensions of the system are negligible in comparison. We can put it another way by saying that the image is so far away, that if we double the distance, nothing changes in the system. Notice, however, that if we move the lens so that the image moves out to infinity, and then keep moving the lens, the image comes into view from the other side, that is, negative infinity is "adjacent to" positive infinity. We find the same behavior with the tangent of an angle, as the angle moves through 90 degrees, so the mathematics corresponds to the physics.

To a mathematician, infinity is a concept of unendingness, not really a magnitude at all. To a physicist it is a countable magnitude, with no built-in limiting value, that is, infinity is indefinitely large. The difference arises when we ask whether being able to do something as often as you like means the same as doing it forever.

Zeno's famous paradoxes show what is involved. Achilles runs after the tortoise, but every time he gets to where the tortoise was, the tortoise has moved on. This goes on forever, so Achilles can never overtake the tortoise, even though he runs a lot faster. Similarly, the arrow never can reach the target because every time it covers half the distance, there is still a distance yet to cover.

There is a serious principle to be drawn from this clever nonsense: extrapolating from a valid finite sequence of operations to an *infinite* sequence requires *additional* justification. Specifying an indefinitely *large* repetition of operations is not at all the same as concluding that the *infinite* result is valid. We can divide a finite real number by smaller and smaller divisors, but we cannot divide by zero. The limiting case is *not* included. We can see what is wrong when we start to count the even integers: we count 1 for 2, 2 for 4, 3 for 6, etc. In one view, we seem to be counting the even integers, but the integers are falling further and further behind the things (even numbers) we are counting, so that "in the limit" we don't count them. In Zeno's paradoxes, each successive position of Achilles or of the arrow makes less and less difference, so "in the limit" they have stopped moving; the condition of motion is violated, and so is Zeno's conclusion. If you want to include the infinite case, you need an *additional* assumption or piece of information.

Imaginary Numbers

We cannot picture, even approximately, what the square root of minus four meters looks like, so it comes as a surprise to most people that imaginary numbers are very useful in physics and engineering. Although the square root of a negative number may seem like an impossibility, an aberration of the law of signs, imaginary numbers are full-fledged, legitimate members of the family of numbers; in fact, they are necessary to complete the family. Theorems about functions of numbers which must be real (not imaginary) are harder to prove, and less general, than when the functions are of complex (real plus imaginary) variables. When an equation has complex terms in it, one can split the equation into two parts, one real, one imaginary, and thus simplify the problem by solving them separately. The two-headed character of complex quantities resembles vectors.

The voltage in an electric circuit is ordinarily expressed in real numbers, but an imaginary term can indicate the phase shift in the circuit. The index of refraction of glass is ordinarily expressed in real numbers, but an imaginary term can give the absorption of light by the glass, that is, gray or colored glass. Quantum mechanics relies heavily on complex vectors; the Schrödinger equation has the square root of minus one built right into it. The four dimensions of relativity theory are most neatly expressed if time is included as an imaginary term.

Real and imaginary quantities are more intimately related than appears at first sight. Consider what it means when two graphs, each represented by an equation, cross each other. The point, or points, or intersection represent common solutions of the two equations. That is to say, the co-ordinates of the points shared by the two equations must be values which will satisfy each of the equations separately. Take the horizontal line $y = 1$, and the diagonal line $x = y$. Solve them simultaneously, and you get the common co-ordinate point: (1,1), which is where they cross. If we had used $y = 1$ and $y = 2$ for the two equations, we would not get any common co-ordinate, because the two lines are parallel, and don't cross anywhere.

We can find the intersection points with more complicated equations, represented in Figure A1.

$$x^2 + y^2 = 4 \quad \text{and}$$
$$x^2 + y^2 = 5$$

These are the equations of two concentric circles, with their common center at the origin. The two circular lines are parallel, and do not intersect. Therefore, we get no common solution to the equations. But now let us take the smaller circle,

$$x^2 + y^2 = 4$$

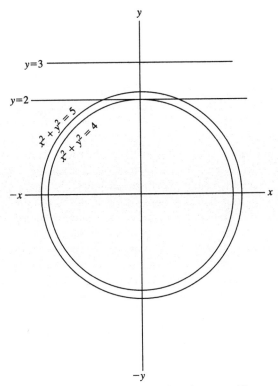

Figure A1   Where do they intersect?

and the horizontal line $y = 3$, which is too high to intersect the circle. Or is it? We get a solution, actually two of them, when we solve the equations simultaneously:

$$(i\sqrt{5}, 3) \quad \text{and} \quad (-i\sqrt{5}, 3)$$

Where are these intersection points? If we took $y = 2$, we would get the tangent point in real space as the intersection. But where is the imaginary space?

The analysis can be extended using other equations, and the imaginary space explored, or mapped. Here we merely point out that we have found a hidden correlation between real and imaginary quantities. They sneak out and meet each other. This "sneaky" quality is occasionally found among physical variables too, such as the phase and voltage of a circuit, or the refraction and absorption of a glass. Heisenberg's uncertainty principle demonstrates that the momentum and position of a moving particle have a hidden correlation. Bell's theorem involves a hidden correlation between certain pairs of separated particles.

As this discussion suggests, the applications of mathematics in physics can be subtle. There are risks and limits, on the physicists' side. Solutions to valid equations may not have physical meaning. Conversely, physical states may exist when the appropriate mathematical equation can be shown to have no solutions. There are cases where we deliberately violate the conditions necessary to render a mathematical theorem valid, and get physically useful results, which check out experimentally.

Some of the discoveries of modern physics consist of little more than mathematical demonstrations which, hopefully, parallel ("explain") the observed physical processes. How far should such demonstrations be trusted? Physicists invent, or adapt, mathematical tools as needed. In physical situations where vectors do not transform as we would like, we define polar vectors, and spinors, instead of axial vectors; we have symmetric and antisymmetric tensors; we have various types of complex (imaginary) quantities, Dirac delta functions, and all manner of ad hoc mathematical creatures. Physicists have the same rights as mathematicians in conjuring up creatures which behave as needed, but questions of physical reality and validity eventually need to be faced. The question of hidden variables forbidden by quantum theory ("foolish questions") is a case in point.

Appendix E
Invariance of Vectors

Vectors are two-headed mathematical quantities. We encountered them in connection with fields. If each point in a field is characterized by two quantities, such as magnetic strength and direction (angular deviation from pointing to the north pole), vectors are the most convenient mathematical tools. But vectors have an additional property, and this endears them particularly when we are concerned with relativity. This second property is their invariance.

In plain English, invariance means that they remain unchanged when we change the coordinate axes. If we note the coordinates, or addresses, of each end of a vector, then the length of the vector is just the length of the straight line between those two points, as shown in Figure A2. But the nice thing is that if we want to indicate the length of the vector using *different* axes, we get the same length, even though the coordinates of each end point have changed.

In Figure A2 we have a vector whose length squared is 13. One end of the vector is at (1,1) and the other is at (4,3). (These coordinates can be read off on the first pair of x-y axes in Figure A2). If we shift our axes by three units to the left along the x direction, and two units down in the y direction, the coordinates for the end points are (4,3) and (7,5). If we shift

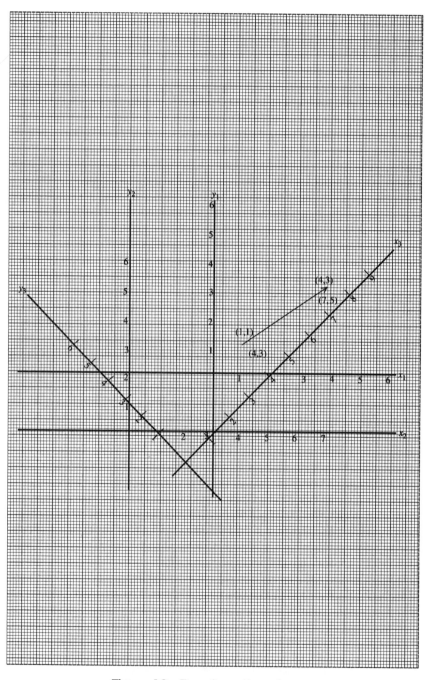

Figure A2   Transformation of axes.

the axes again, this time also rotating them by 45 degrees, the coordinates for the end points are approximately (4.2, 1.41) and (7.74, 0.71).

The three calculations of the length of the vector, one for each pair of coordinate addresses, are each given by the Pythagorean theorem. The detailed justification for this is given in textbooks on analytic geometry.

$$d^2 = (4 - 1)^2 + (3 - 1)^2 = 3^2 + 2^2 = 13$$
$$d^2 = (7 - 4)^2 + (5 - 3)^2 = 13$$
$$d^2 = (7.74 - 4.20)^2 + (0.71 - 1.41)^2 = 3.54^2 + 0.7^2 = 13$$

Invariance is important in physics because most physical processes are independent of where they occur. An experiment you do in your laboratory should work the same in my laboratory. So where we put the coordinate axes, and which direction they face, shouldn't matter. The example given in Figure A2 is restricted to two dimensions, the plane of the paper, but the invariance property of vectors holds for three dimensions, or four, or any number of dimensions. Better yet, we can extend the idea of an invariant two-headed number to include three-headed numbers, and four, etc. These multiheaded monsters are called tensors. A vector is a tensor of rank 1; an ordinary scalar number (e.g., number of apples or degrees of temperature) is a tensor of rank zero. The general relativity theory uses tensors of rank 2 and rank 3, which are horrible.

Appendix F

Derivation of Snell's Law from Geodesy Principle (Fermat's Principle)

Figure A3 shows a light ray going from a starting point to a finishing point, being refracted as it passes into the medium in which it is propagated more slowly. We wish to find the condition (i.e., the distance x), which allows the ray to complete the trip in shortest time. The trip, or trajectory time, is given by adding the time in each medium. These times are found by dividing the distance (which forms the hypotenuse of a triangle in the figure) by the velocity in that medium. We use the Pythagorean theorem, and get

$$\text{Trajectory time} = \frac{\text{distance in faster medium}}{v_1}$$

$$+ \frac{\text{distance in slower medium}}{v_2}$$

$$T = \frac{\sqrt{a^2 + x^2}}{v_1} + \frac{\sqrt{(L - x)^2 + b^2}}{v_2}.$$

The geodetic principle (in this case the same as Fermat's principle) is:

$$\delta \int_{t_0}^{t_1} T \, dx = 0,$$

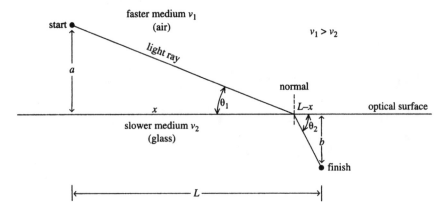

Figure A3   Diagram of a light ray refracted at a glass surface.

where t_0 is the starting time for the light and t_1 is the finish time. Thus, we have for the integral

$$\int \left[\frac{1}{v_1} (a^2 + x^2)^{1/2} + \frac{1}{v_2} ((L - x)^2 + b^2)^{1/2} \right] dx.$$

Since the integrand contains no derivative (which is why I chose this case), we can find its stationary value simply by setting $dT/dx = 0$:

$$\frac{dT}{dx} = \frac{1}{2v_1} (a^2 + x^2)^{-1/2} \cdot 2x + \frac{1}{2v_2} \{(L$$

$$- x)^2 + b^2\}^{-1/2} \cdot - 2(L - x) = 0,$$

whence

$$\frac{2x}{2v_1 \sqrt{a^2 + x^2}} = \frac{2(L - x)}{2v_2 \sqrt{(L - x)^2 + b^2}}.$$

Using the angles in Figure A3, we convert to cosines:

$$\frac{\cos \theta_1}{v_1} = \frac{\cos \theta_2}{v_2}.$$

The angles, θ_1 and θ_2, are measured between the ray and the surface. The convention in optics is to use the angles between the ray and the normal to the surface, unlabeled in Figure A3. These angles are known as the angle of incidence, i, and the angle of refraction, r. Thus we have

$$\frac{\sin i}{\sin r} = \frac{v_1}{v_2},$$

which is Snell's law.

A more elaborate analysis would allow us to use more than 2 media. By reducing the ray lengths in many media toward the vanishing point, we can represent refraction in media with continuously varying refractive index. In such cases, the ray trajectory is usually curved. Feynman[1] gives a geometric demonstration of the situation in his *Lectures on Physics*, together with a similar derivation of the law of reflection from Fermat's principle.

The length of a curve in two-dimensional space, from a to b, is

$$L(a, b) = \int_a^b \sqrt{1 + \left(\frac{dy}{dx}\right)^2}\, dx,$$

assuming that the derivative exists and is continuous. In a system space of n dimensions, this trajectory becomes

$$L(a, b) = \int_a^b \sqrt{\left(\frac{dy_1}{dx}\right)^2 + \left(\frac{dy_2}{dx}\right)^2 + \cdots \left(\frac{dy_n}{dx}\right)^2}\, dx$$

$$= \int_a^b \sqrt{\sum_i^n \left(\frac{dy_i}{dx}\right)^2}\, dx.$$

The geodetic principle then can be written:

$$\delta \int L(a, b) = 0.$$

This involves a double integration, because L is already an integral. The calculation becomes even worse if we require, instead of a *line* trajectory, that an *area* or *volume* be an extremum. The mathematics extends beyond the scope of this book.

---

[1] R. Feynman, *Lectures on Physics*. Addison-Wesley. 1963. Vol. I-26.

Index

Page numbers followed by an "f" represent figures; a "t" following a page number indicates a table.